ROOF COOLING TECHNIQUES A DESIGN HANDBOOK

Supplementary Resources Disclaimer

Additional resources were previously made available for this title on CD. However, as CD has become a less accessible format, all resources have been moved to a more convenient online download option.

You can find these resources available here: www.routledge.com/9781844073139

Please note: Where this title mentions the associated disc, please use the downloadable resources instead.

ROOF COOLING TECHNIQUES A DESIGN HANDBOOK

Simos Yannas | Evyatar Erell | Jose Luis Molina

earthscan
from Routledge

First published by Earthscan in the UK and USA in 2006

For a full list of publications please contact:
Earthscan
2 Park Square, Milton Park, Abingdon, Oxon OX14 4RN
52 Vanderbilt Avenue, New York, NY 10017 USA

Earthscan is an imprint of the Taylor & Francis Group, an informa business

A catalogue record for this book is available from the British Library

Library of Congress Cataloging-in-Publication Data

A catalog record for this book has been applied for

Cover design by Paul Cooper Design

ISBN 13: 978-1-84407-313-9 (pbk)

CONTENTS

PREFACE

This design handbook of roof cooling techniques is structured in two main parts. Part I deals with principles of roof design. Chapter 1 divides the environmental functions of roofs into protective (solar control, thermal insulation, heat storage and thermal capacity) and selective (encompassing radiative, evaporative and convective cooling, as well as planting of roofs) introducing the main physical principles involved in each of these. Chapter 2 provides an overview of traditional and current construction techniques, drawing upon examples from several European countries. Part II of the Handbook focuses on specific systems and techniques. Chapters 3, 4 and 5 focus respectively on roof ponds, cooling radiators (with water or air as heat transfer fluid) and planted ("green") roofs. Each of these chapters provides an exposition of the fundamental principles of the respective technique, the systems and components employed, advantages and disadvantages, application types, built precedents and recent experimental work (where applicable), and key design considerations. Applicability and performance are then dealt with in more detail in Chapter 6, which can be used by readers to make choices on system selection and design. The data and design guidelines given in Chapter 6 are also intended to provide comparative performance indicators for readers who wish to use the software included with the Handbook so that the results they produce can be assessed against these data. Both air-conditioned and free-running buildings are covered by the data and guidelines, and applications of roof ponds are considered on both single-storey and multi-storey buildings. The constructional specifications encompass both lightweight and heavyweight roof types with different levels of thermal insulation.

The Handbook includes an extensive bibliography that covers all the main scientific and technical aspects of the topics. There are two appendices: Appendix A presents the mathematical models that were developed for the study of the roof cooling techniques encompassed by the Handbook. These are implemented in the design support software which is included with this Handbook and introduced in Appendix B. The software will allow readers of the Handbook to undertake evaluations of their own designs. Its database of weather data covers a number of locations and users can produce weather files for other locations using the software's file format. Building form is identified graphically, whilst constructional specifications of building elements and the operational characteristics of different cooling techniques can be selected from the menus provided. Calculations can be performed for a single location or to map applicability within a given geograpic region. For each cooling technique modelled by the user the tool also creates an air-conditioned or free-running reference building case against which the cooling technique is assessed in terms of possible cooling energy savings and/or improvements in indoor thermal comfort conditions.

ACKNOWLEDGMENTS

This book and the accompanying software were produced in the context of a project carried out within the framework of the ALTENER Programme of the European Commission Directorate General for Energy and coordinated as part of a clustered project, Cluster 9, by WSP Environmental, London, UK. Much of the material for the book originates in research carried out by the authors and others in the context of a project on Roof Solutions for Natural Cooling (ROOFSOL) that was partially supported by the European Commission DG XII under the Joule Programme. We would like to acknowledge the contributions from the many colleagues who took part in various stages of these projects. In particular: Alejandro Branger, Gustavo Brunelli and Helena Massa of the Environment & Energy Studies Programme, Architectural Association Graduate School, London, for their contribution in literature research and production of illustrations; Eugenia Lazari, of the Centre for Renewable Energy Sources (CRES), Athens, Helena Granados, consultant architect on ROOFSOL for CIEMAT, Spain, F. Aleo, S. Scalia and G. Panno of Conphoebus, Catania for reviews and compilations of illustrations of traditional and contemporary roof constructions in Greece, Spain and Italy respectively that provided material for Chapter 2 of this book; Prof. Yair Etzion for experimental work reported in Chapters 3 and 4 and performed at the Center for Desert Architecture and Urban Planning (CDAUP), Sde-Boqer; Felix M. Tellez and Guillermo Schwartz, who also coordinated the ROOFSOL project on behalf of the Centro de Investigaciones Energéticas Medioambientales y Tecnológicas (CIEMAT), Madrid for experimental work and illustrations on roof ponds and planted roofs that are summarised in Chapters 3 and 5 of this book respectively; Dr Argiro Dimoudi, Gordon Sutherland, Andreas Androutsopoulos and M. Vallindras for the experimental work on a water radiator system that was monitored by them at the Centre for Renewable Energy Sources (CRES), Athens and is discussed in Chapter 4; Prof. Eduardo Rodriguez, University of Malaga, Spain for his contribution to the mathematical models summarised in Appendix A and implemented in the RSPT software and for the parametric studies on radiators summarised here in Chapter 4; Prof. Elena Palomo Del Barrio, presently of the University of Bordeaux, France for her analysis of planted roofs that is summarised in the last section of Chapter 5 and in its mathematical form in Appendix A; Dr José Manuel Salmerón Lissén and his collaborators at Departamento de Termotecnia, Escuela Técnica Superior de Ingenieros Industriales, Universidad de Sevilla for running the parametric studies summarised in Chapter 6 and for tireless revisions of the RSPT software described in Appendix B and included on CD with this publication. We also acknowledge the contributions from Prof. Gilles Lefebvre and collaborators from GISE, Ecole Nationale des Ponts et Chaussées, Paris for work on modelling of roof cooling systems; Prof. Servando Alvarez, School of Engineering, University of Seville for his valuable contributions at various stages of the project; Harold Hay for his pioneering work on roof ponds.

Photograph and Illustration Credits

1.2 Chetwood Associates Architects, London.
1.11 Kirsten Davis, Architectural Association Graduate School, London.
1.12 Reproduced from P. Berdahl and R. Fromberg, Solar Energy Vol.29, No. 4, Pergamon Press Ltd., 1982.
1.17 Spreadsheet software by D. Robinson with weather data generated by Meteonorm v4.0 (Meteotest).
1.20 & 1.22 Benito Jimenez Alcala, Architectural Association Graduate School, London.
2.1-2.3 Dimitri Philippides, Greek Traditional Architecture, Melissa Publishers, 1984.

2.14 & 2.16 (top left) M. Papadopoulos, Thermal Insulation of Buildings, Monyal 1979.
3.5 (bottom) Margot McDonald, Cal Poly, USA.
3.5 (top), 3.6, 3.8 & 3.9 Reproduced from W.P. Marlatt, K.A. Murray, S.E. Squir, Roof Pond Systems, Energy Technology Engineering Centre ETEC-83-6, 1984, Calif. USA.
5.3 Hockerton Housing Project, UK.
5.7-5.9 E. Georgiadou, architect, Thessaloniki, Greece.

All other photographs and illustrations by the authors or as credited in the Acknowledgments above.

1 ENVIRONMENTAL FUNCTIONS OF ROOFS

INTRODUCTION

Roofs offer protection from the elements, but can also help exploit ambient energy sources and sinks, contributing to the space heating, cooling, ventilation and daylighting of buildings (Figs 1.1–1.2). The mainstream approach to roof design has emphasised the roof's *protective* function. This includes protection from sun, ambient temperature, wind, rain and snow. In most European countries building regulations enforce a prescriptive application of thermal insulation as a means of reducing the use of energy for space heating and cooling.

Though almost always necessary, protective mechanisms are not sufficient in themselves to make a building independent from conventional heating and cooling. To acquire such independence whilst achieving thermal comfort conditions for its occupants, a building requires suitable energy sinks for the dissipation of excess heat as well as renewable sources for space heating at other times. The latter are readily available from sunshine and from a building's internal heat gains due to occupancy. The permanent heat sinks are the ambient air, the ground, water masses and the sky when their temperatures are suitably lower than those of the spaces we aim to cool.

In this book we investigate the *selective* coupling of two of these heat sinks, the ambient air and sky, with roof elements aiming to provide heat dissipation and cooling for occupant thermal comfort. Roofs are generally the most exposed element of a building's external envelope. The balance between the protective and selective environmental functions of the roof is a function of temporal, as well as contextual parameters. The following topics are introduced in this chapter.

Protective functions
• solar control
• thermal insulation
• heat storage

Selective functions
• radiative cooling
• evaporative cooling
• convective cooling
• roof planting

Fig. 1.1 Partially transmissive awning as movable protective and selective device on courtyard glazing in Seville, Spain.

Fig. 1.2 Selective opening of the roof for daylighting on a commercial building, London, UK.
(Architect: Chetwood Associates, London)

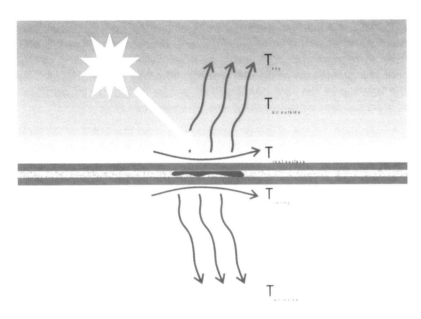

Fig. 1.3 **Solar and thermal processes on external and internal surfaces of a flat roof.**

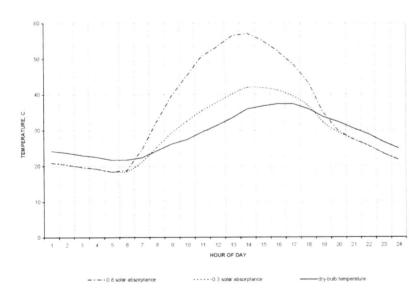

Fig. 1.4 **Hourly ambient air temperature and temperatures of external horizontal surfaces of 0.3 (low) and 0.8 (high) solar absorptance on a typical summer day in Athens, Greece.**

PROTECTIVE ENVIRONMENTAL FUNCTIONS

Solar Control

The roof is commonly the element of a building that is most exposed to the sun. Solar radiation absorbed by roof surfaces raises the surface temperatures, driving heat transfers toward the interior of buildings, as well as towards the ambient air and sky (Fig. 1.3). The graph in Fig. 1.4 illustrates this by comparing the values of the air temperature on a typical summer day in Athens, Greece (obtained using the Meteonorm software**) with the temperatures of horizontal surfaces of solar absorptance 0.30 and 0.80, respectively*. It can be seen that the rise in the surface temperatures above that of the ambient air is a function of surface solar absorptance. The peaks in surface temperatures are affected by solar radiation which peaks at noon on horizontal surfaces, whereas the outdoor air temperature has its peak in the early afternoon hours. The drop in surface temperatures below the ambient air at night reflects the effect of radiative cooling, i.e. a net heat loss by longwave radiation to the night sky.

Clearly, the design of a roof should seek to control the absorption of solar radiation and its effect indoors. This may be achieved by one or more of the following:

- suitable choice of orientation, tilt angle and surface area of roofs; these parameters affect the amount of incident solar radiation;
- light-coloured external finishing to reduce absorptance of solar radiation;
- shading of the roof to reduce incidence of direct solar radiation.

There are trade-offs between some of these parameters. For example, because of higher exposed surface area, a vaulted roof may receive more solar radiation than a flat roof in summer, but will also incur higher heat losses because of its higher exposure. It has been shown that the net difference may be quite small (Pearlmutter, 1993). A pitched roof tilted away from the sun receives less direct radiation than a horizontal roof; however, its larger surface area also results in greater exposure to diffuse radiation. The effect of surface colour on the absorption of solar energy is more clear-cut. Suitably maintained whitewashed roofs, whether flat or vaulted, have long been a feature of some Mediterranean cities because of their ability to reflect as much as 80% of the incident solar radiation

* solar absorptance is the ratio of radiation absorbed at a surface to the radiation incident on that surface. For an opaque surface absorptance + reflectance = 1.

** Meteonorm v5.0 (2003). Global Meteorological database for solar energy and applied climatology, Meteotest, Bern.

Fig. 1.5   **Vaulted and whitewashed roofs are a characteristic of the traditional architecture on the island of Thira (Santorini), Greece.**

Fig. 1.6   **Diffusing glazing reduces solar transmission and provides a more even distribution of daylight, combining protective and selective strategies, Musee d'Orsay, Paris.**

(Fig. 1.5). Attention to the solar transmissivity of roof openings and awnings is essential to avoid glare and overheating (Figs 1.6 and 1.7). In regions with high levels of solar radiation, shading of the roof can provide an additional strategy. Shading can be provided by the layering of the roof construction, and/or by the use of additional external elements including use of vegetation (Fig. 1.8).

An indication of the effect of solar radiation on the *external surface* of an opaque building element is given by the *sol-air temperature,* an approximation of the external surface temperature derived from the following simplified expression.

$$t_{sa} = t_o + \frac{\alpha I_t - \varepsilon \Delta R}{h_o}$$

where,

t_{sa},   sol-air temperature, °C;
t_o,   ambient air temperature, °C;
α,   surface solar absorptance (dimensionless);
I_t,   total solar radiation incident on surface, W/m²;
h_o,   external surface heat transfer coefficient*, W/m²K;
ε,   surface emittance (dimensionless);
ΔR,   net radiative exchange (difference between longwave radiation; received and emitted by surface), W/m².

As an example, consider a flat roof covered with grey tiles and exposed to an ambient air temperature of 35°C and incident global solar radiation of 800 W/m². These are typical southern European conditions for a clear summer day around noon. The values of solar absorptance may vary from under 0.20 for very light-coloured, clean surfaces to over 0.90 for very dark or dirty surfaces. Assuming a solar absorptance of 0.70 for the grey tiles, a value of 25 W/m²K for the external surface heat transfer coefficient h_o (based on a wind speed of 3 m/s at roof level)** and no net exchange by longwave radiation at that time, the sol-air temperature at the external surface of this roof can be estimated as:

$$t_{sa} = 35 + 0.70 \times 800 / 25 = 35 + 22.4 = 57.4°C$$

This compares with values around midday on the graph of Fig. 1.4. In city centres, where wind speeds are low, the external heat trans-

Fig. 1.7 The white PVC covering on the Palenque at the site of the World Expo in Seville has an external solar absorptance of 0.10, reflectance of 0.77 and transmittance of 0.13%; the material was wetted externally to lower the surface temperature by evaporation.

Fig. 1.8 Shading effect of vegetation on roof.

fer coefficient h_o will take lower values than 25 W/m²K; for example, h_o =14 W/m²K for sheltered conditions in inner city areas**. As a result, the value of t_{sa} can rise well above the value predicted above. Dirt and air pollution make surfaces darker, increasing their solar

Thermal Insulation

The thermal exchanges between buildings and the outdoor environment depend on the temperature difference between inside and outside, as well as on the exposure and thermal properties of external building elements. The application of thermal insulation suppresses these exchanges. The larger the temperature differences between inside and outside, the more important thermal insulation becomes in reducing the use of conventional energy sources for space heating and cooling. Similarly, its importance increases the more extreme the climate and/or the more exposed a building's envelope is to outdoor conditions. The maps in Fig. 1.9 illustrate the effect of the latter on residential demand for space heating and cooling across parts of Europe. Whilst in northern Europe the demand for space heating is much higher than that for cooling, space cooling has been increasing in importance, especially in southern regions and in urban centres.

Thermal insulation materials are commonly of very low density (10–30 kg/m³) and of low thermal conductivity (typically, in the range 0.025–0.040 W/m²K). The application of such materials is appropriate where the dominant mode of heat transfer across a building element is conduction. Typical thicknesses of insulation layers may vary between 25 mm and 300 mm depending on the temperature difference for which the insulation is applied. Where heat transfer is by longwave radiation, as for example in cavities, the appropriate materials for suppressing it are those of low thermal emissivity*, such as aluminium foil and the metallic coatings applied to glazing.

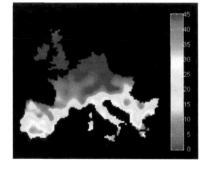

Fig. 1.9 Calculated annual demand in kWh/m² dwelling floor area for residential space heating (left) and cooling (right) in Europe.

* **Emissivity** is a measure of the ability of a surface to emit longwave radiation relative to a *blackbody*, a perfect radiator with an emissivity of 1.0. Most common building materials have a high emissivity (in the region of 0.90). Polished metals and specially coated surfaces can achieve low emissivity values of 0.02–0.05.

Heat Storage and Thermal Capacity

Heat is stored in the mass of the walls, partitions, floors and ceilings of buildings. Heat storage in building elements acts as an interim heat sink that helps stabilise internal temperatures at times of fluctuation. In the cooling season this is of value in all cases where diurnal temperature fluctuations caused by external factors or occupant activity could lead to overheating and thermal discomfort. We measure the ability of a building element to store heat by its thermal capacity. This is the product of an element's density (in units of kg/m³) by its volume (in m³) and specific heat (in J/kgK or Wh/kgK) and has units of J/K or Wh/K. It is a measure of energy stored per degree temperature rise in the material. A free-running* building with a low thermal capacity tends to follow the fluctuations of the outdoor air temperature and those caused by internal heat gains from occupancy and incident solar radiation. The latter lead to rapid temperature rises above that of the outdoor air which in itself may be too high for occupant comfort at times. Similarly, when the outdoor air temperature falls it is followed by a rapid drop in indoor temperature that may also lead to occupant discomfort. At the other extreme, a building of very high thermal capacity tends towards constant room temperatures. In Europe traditional architecture produced buildings of fairly high thermal capacity. This was a direct result of the use of high density masonry materials for external walls and floors. In warm weather this had a positive effect in keeping indoor temperatures below outdoor peaks during daytime in summer. An extreme form of high thermal capacity construction evolved on the island of Santorini in Greece (Fig. 1.10). This is a partly earth-sheltered construction built of the island's lava stone and volcanic soil (see also Chapter 2). A contemporary residential scheme that makes use of the thermal capacity of the earth for insulation and thermal inertia is the group of five single-storey housing units at Hockerton in England (Fig. 1.11). The northern side of the roof is covered by 400 mm of topsoil over a roof construction of reinforced concrete beams and blocks insulated with 300 mm of expanded polystyrene.

In many countries contemporary construction relies on reinforced concrete slabs and masonry external walls. This combination can provide a considerable thermal capacity if surfaces are left exposed. However, floors and walls are frequently covered with carpets, furniture and other lightweight finishes, leaving ceilings as the only exposed surfaces for heat storage.

* A building is free-running when not mechanically controlled by a heating or cooling appliance.

Fig. 1.10 Earth-sheltered masonry building on Santorini, Greece.

Fig. 1.11 View from the north of earth-sheltered terrace of houses at Hockerton, Nottinghamshire, UK.

SELECTIVE ENVIRONMENTAL FUNCTIONS OF ROOFS

Radiative Cooling

The sky is the ultimate heat sink and longwave radiation the main mechanism by which surfaces on Earth dissipate the heat absorbed from the sun on a daily basis. During daytime, building surfaces absorb more heat than they can dissipate by longwave emission. After sunset, these surfaces may be still exposed to longwave radiation from the atmosphere but, owing to the absence of solar gains, a net cooling effect can be achieved. This is a function of the difference between the incoming and the outgoing longwave radiation and depends on local climatic conditions. With a clear sky and low humidity, the *incoming* longwave radiation from the sky is concentrated in two distinct wavelength bands, 3 to 8 μm and 13 to 20 μm. In the range of 8 to 13 μm, the atmospheric radiation is at its weakest. This wavelength band is known as the *atmospheric window* for the radiative cooling of terrestrial surfaces. The atmospheric window is particularly noticeable near the zenith of the sky. It is also strongly affected by cloudiness and humidity. This is because water vapour in the air absorbs outgoing longwave radiation, reducing the net rate of heat dissipation. Clouds have an analogous effect and overcast skies inhibit net radiative cooling. The graphs in Fig. 1.12 illustrate these points. The graph on the left shows how the radiation emitted by the sky decreases towards the zenith (zenith angle of 0) thus highlighting the opening of the atmospheric window in this region. The graph on the right illustrates the same effect as a function of water vapour in the atmosphere, showing how reduction in water vapour results in lower sky radiance.

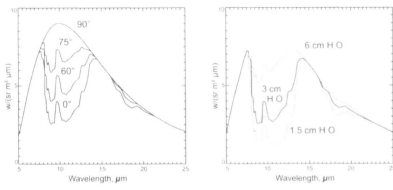

Fig. 1.12 Spectral radiance of sky as a function of zenith angle (left), and spectral radiance (zenith) as a function of water vapour in the atmosphere (right) showing sensitivity of the atmospheric window. (Source: Clark, E., 1981)

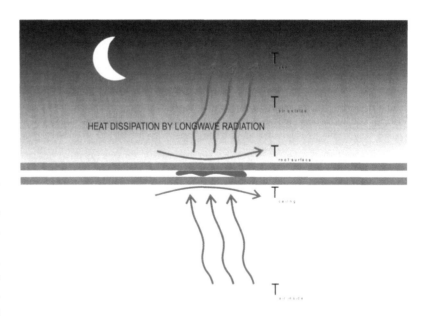

Fig. 1.13   Thermal exchanges of roof surfaces at night-time.

The only building element that can be used for radiative cooling is the roof because of its view of the sky zenith. Horizontal roofs have a better view of the zenith region than pitched roofs. Longwave radiation emitted by building surfaces is commonly within a spectral band 5–30 μm with a peak wavelength at approximately 10 μm.

The radiant energy exchange between a building surface and the sky (Fig. 1.13) is given (in W/m²) by:

$$Q = \varepsilon_b\, \sigma\, (T_b^4 - T_{sky}^4)$$

where,
$\sigma = 5.67 \times 10^{-8}$ W/m²/K⁴ is the Stefan-Boltzmann constant;
ε_b and T_b are the emissivity (dimensionless) and (absolute) temperature of building surfaces in Kelvin;
T_{sky} is the sky temperature*, K.

Most building materials have a high emissivity and are thus good radiators. Assuming a value of emissivity of 0.9, surface temperatures of 303K (27°C) and sky temperature of 288K (15°C), this expression gives a cooling output of 79 W/m² of radiator surface.

*__Sky temperature__ is the temperature of a uniform hypothetical black body surface that would give off the same radiant flux as the atmosphere.

The larger the area of radiator the larger the amount of heat that can be dissipated (Fig.1.14). The values of sky temperature depend on atmospheric conditions and on the ambient air temperature. Figure 1.15 compares hourly values of sky temperature to those of ambient air temperature on a sunny summer day in Seville, Spain. Note how the difference between air and sky temperatures, the *sky temperature depression**, increases at night reaching its maximum when the air temperature falls to its daily minimum value. The sky temperature depression provides an indication of the cooling potential by radiation. The radiative cooling potential is highest when sky temperatures are at their lowest; this occurs with clear skies. Figure 1.16 shows contours of night-time sky temperature depression (between 2200 and 0600 hours) for summertime in different parts of Europe.

Sky temperature depression is the difference between the sky temperature and the ambient air temperature.

For small differences between the temperature of a radiating surface and the sky temperature, the following form of the Stefan–Boltzmann Law may be used to estimate the net radiative heat loss in W/m^2 from the surface of the radiator (Martin, 1989):

$$R_{net} = 4\varepsilon\sigma\, T_{air}^{\;3}(T_{rad} - T_{sky})$$

where,
R_{net} net radiative heat loss, W/m^2;
ε emissivity of the radiator surface;
σ Stefan–Boltzmann constant;
T_{air} ambient air temperature, K;
T_{rad} radiator temperature, K;
T_{sky} sky temperature, K.

For the range of ambient air temperatures normally encountered in regions where radiative cooling may be applied, typically 15–30°C (288–303K), the expression $(4\varepsilon\sigma\, T_{air}^{\;3})$ results in values around 5 W/m^{2}K. For this value the net cooling output of a radiator is approximately five times the difference between the radiator and sky temperatures. It can be seen from this that the higher the temperature of the radiator the larger the cooling output.

External building surfaces acting as radiators will also enter into thermal exchanges by convection with the surrounding ambient air. The effect of convection is directly proportional to the temperature difference between the radiator and the ambient air. This is shown by the following expression.

Fig. 1.14 Commercially available polypropylene radiator.
(See Chapter 4 for test results)

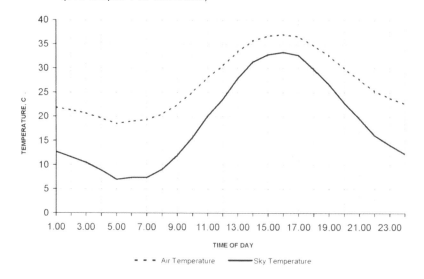

Fig. 1.15   Hourly values of sky temperature and ambient air temperature for a summer day in Seville.

Fig. 1.16   Contours of sky temperature depression (in units of degree-hours).

$$Q_c = h_c (T_r - T_a)$$

where,

Q_c rate of convective heat exchange, W/m^2;

h_c convective heat transfer coefficient, W/m^{2}K;

T_r, T_a temperature of radiator and ambient air, $^\circ$C.

For as long as the radiating surfaces are at higher temperatures than the air, this process assists in heat dissipation. However, because radiating surfaces are dissipating heat to the cooler sky their temperatures can fall below those of adjacent air layers. When this happens the convective heat transfer acts as a heat gain, warming up the radiating surfaces towards the ambient air temperature. It is thus desirable that the radiator should remain warm in line with the reduction in temperature of indoor spaces.

The value of the convective heat transfer coefficient h_c depends on wind speed and on the relative temperatures of the radiator and the air (Clark and Berdahl, 1980). Typical values of h_c may range from under 3.0 W/m^{2}K to above 10 W/m^{2}K. For a radiator temperature T_r of 30°C and an ambient air temperature T_a of 25°C, the resulting heat transfer by convection may then vary from around 15 W/m^{2}K to over 50 W/m^2. It can be seen from the above that where climatic conditions are suitable, careful design can result in a radiative system capable of providing a net cooling output on the order of 100 W/m^2 of installed radiator surface, for most of the night.

Design guidelines on radiative cooling are given in Chapters 3, 4 and 6.

Evaporative Cooling

The evaporation of water, a phase change from liquid to gas (vapour), is driven by the absorption of heat from the surrounding air. The air in contact with this process loses heat and cools but gains moisture, becoming more humid. The energy required for the phase change of water from liquid to gas is known as the *latent heat of vaporisation*. When this is taken from the surrounding air the process is known as *adiabatic*. The relationship between the temperature and moisture content of the air is presented graphically on the psychrometric chart shown in Fig. 1.17. The absolute humidity or humidity ratio is expressed in grams of water per kilogram of dry air on the vertical axis. The curved lines represent the

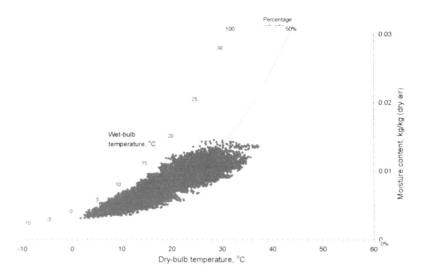

Fig. 1.17 Psychrometric chart with plot of concident hourly dry-bulb and relative humidity values for summer (weather data for Athens).

relative humidity, that is a percentage measure of saturation. The temperature of air saturated with moisture is the wet-bulb temperature; its scale is read on the line of 100% relative humidity. The dry-bulb corresponding to saturation is the dew point temperature. The amount of energy required to evaporate water at a temperature of 25°C under standard atmospheric pressure of 100 kPa is 2.44 MJ/kg. Under these conditions, the evaporation of just 1 litre of water could cool approximately 200 m^3 of air by 10°C. With evaporation, air can be cooled to a temperature that is within 2°C of its wet-bulb temperature. Humidity ratios above 12 g/kg are generally considered undesirable, causing discomfort as well as having adverse effects on furniture and building surfaces.

A measure of the potential for evaporative cooling is the *wet-bulb temperature depression*. This is the difference between the dry-bulb and wet-bulb temperatures. Figure 1.18 shows the daily profile of this parameter for a typical summer day in Athens. Unlike the radiative cooling potential which peaks at night, evaporative cooling has its peak during the daytime whilst continuing to have a useful potential at night. Figure 1.19 illustrates the variation of 24-hour degree hours of wet-bulb depression for the summer period across parts of Europe. It can be seen that the largest potential is in the Mediterranean region. Clearly, the application of evaporative cooling is most appropriate in hot and dry climates.

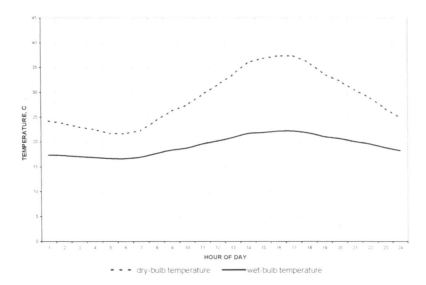

Fig. 1.18 Hourly profile of wet-bulb temperature in summer, Athens.

Fig. 1.20 Traditional applications of evaporative cooling.

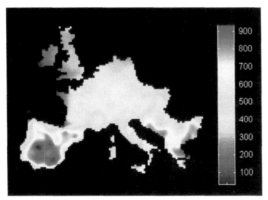

Fig. 1.19 Degree hours of wet-bulb temperature depression over the summer period in Europe.

Evaporative cooling may be used to cool outdoor air supplied to a building for ventilation, or it may be used to cool the building structure. Roofs are amenable to both methods. They are exposed to intense solar radiation, and require a means of reducing surface temperatures. On the other hand, they are exposed to fresh air that may be cooled by evaporation of water and deflected into the building interior, or used to dissipate heat from the roof to ambient air.

Fig. 1.21 Evaporative cooling techniques applied at the site of the 1992 World Expo in Seville to improve the outdoor microclimate.

Techniques based on these processes have formed part of vernacular traditions of cooling buildings and outdoor spaces in hot dry climates (Fig. 1.20). Recent successful applications at the site of the 1992 World Expo in Seville, Spain (Fig. 1.21), have provided quantitative information on their applicability and performance (Macho *et al.*, 1994; Guerra *et al.*, 1994). Cooling of building roofs by evaporation may take two generic forms: roof ponds, in which a mass of water cooled by evaporation and/or longwave radiation is supported on an essentially flat roof (Fig. 1.22); and spray or trickle systems in which water is supplied as needed to cool exterior building surfaces exposed to a hot dry environment (Fig. 1.23).

The cooling of supply air through use of cooling towers that rely on thermal buoyancy and wind forces (passive downdraft evaporative cooling, PDEC) is another approach that has been traditional in hot-dry climates. This has also found successful recent applications on buildings and outdoor spaces in several countries (Figs 1.24–1.25) (Alvarez *et al.*, 1991; Pearlmutter *et al.*, 1996; Francis and Ford, 1999).

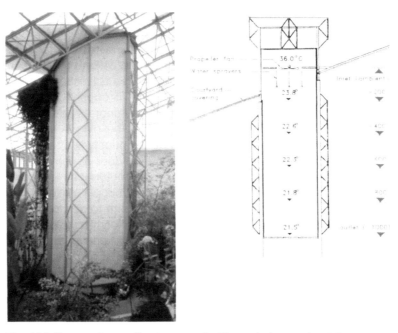

Fig. 1.24 Evaporative cooling tower at the Blaustein International Centre, Sde Boqer, Israel.
(Architects: Centre for Desert Architecture and Urban Planning)

Fig. 1.22 Open roof pond.

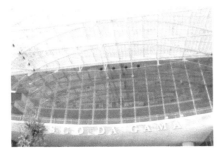 

Fig. 1.23 Vasco da Gama Centre Lisbon. Glazed roof cooled by water running continuously from its ridge.
(Architects: BDP and Promontorio Arquitectos)

Fig. 1.25 Passive downdraft evaporative cooling towers at the site of the 1992 World Expo, Seville.

Roof Openings, Ventilation and Convective Cooling

Ventilation is the provision of a fresh air supply necessary for occupant health and hygiene in buildings. The ventilation process consists of a rate of air exchange that can vary as a function of fresh air requirement, as well as the mechanism of air supply. Depending on the relative temperatures of indoor and outdoor air this process may represent a heat loss or heat gain for the building. A higher air exchange rate than that required for fresh air supply is often aimed at as part of a cooling strategy. When the outdoor air temperature is above the comfort range the introduction of outdoor air to a building at a rate above that required for fresh air supply makes sense only if the resulting air movement can contribute to occupant thermal comfort. This direct contribution to thermal comfort is known as comfort cooling. In hot and dry conditions the benefits of comfort cooling are better realised through the provision of air movement by mechanical means, such as ceiling fans. On the other hand, at times when the outdoor air is pleasantly cool while indoor spaces are too warm for comfort, deliberate flushing of spaces can cool the building structure as well as providing cooler air for occupants. In climates with hot summers this condition occurs only at night.

Ventilation through roof openings may be driven by wind or thermal buoyancy (stack effect; Fig. 1.26–1.29). Pressure differentials resulting from wind are generally much greater than those resulting from the stack effect, so ventilation based on openings oriented to the wind is generally capable of creating a much higher air change rate. On a roof, pressure depends strongly on the slope. On flat roofs, the pressure is generally slightly negative (i.e. causing suction). For steeply sloping roofs, the pressure coefficient is weakly positive on the windward side but negative within the zone of separation on the lee side (Fig. 1.27). The stack effect, on which chimneys operate, is the result of differences in air density between the interior and exterior or between two separate zones in a building. As a volume of air warms up it expands, acquiring a lower density which causes it to rise. In order for a substantial flow of air to occur, the driving force must be strong enough to overcome internal friction and resistance forces caused by contact with the surroundings. The stack effect is limited to situations where there are substantial differences in temperature between the interior and exterior, and where there is a large difference in height between the inlet and outlet. A sufficiently large temperature difference is commonly found in cold climates, where building interiors can be much warmer than the ambient air*. In warm climates, there is seldom a sufficiently large temperature difference, except

* The temperature difference between indoors and outdoors varies continuously, being in the range of 5–15K in mild climates and 15–30K in cooler climates.

Fig. 1.26 Ventilation towers at Queens Building, Leicester, UK.
(Architects: Short Ford & Associates)

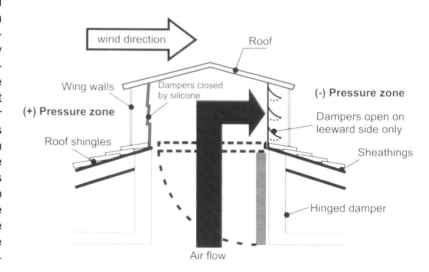

Fig. 1.27 Pressure differences and ventilation through roof openings.

Fig. 1.28 Roof openings for daylight and ventilation at Powergen HQ, Coventry UK. (Architect: Bennett Associates)

Fig. 1.29 Roof openings for ventilation at Cambridge University Department of Mathematics, UK. (Architect: E. Cullinan)

near strong local sources of heat, or at night-time, when the outdoor air temperature falls significantly. The physical processes that affect airflow through buildings are complex. The basic processes may be described by classical fluid dynamics under well-defined conditions. However, real buildings rarely conform to the simplified conditions that may be modelled easily. A variety of methods have been proposed to estimate the air flow rate in a building; a detailed description of prediction methods is given in Allard (1998).

There can be several advantages in employing roof openings for convective cooling:

• wind velocity increases with height above ground, so properly oriented roof openings can introduce a larger flow rate than openings near the ground;
• roof openings may be oriented to wind blowing from any direction; however, the pitch of the roof affects airflow and roof openings may often be in the suction zone;
• roof openings are less likely to be obstructed by neighbouring buildings or trees;
• roof openings are more likely to provide the vertical difference required for stack ventilation;

• in dry, dust-prone environments, elevation of openings above the ground reduces exposure to airborne dust;
• in deep-plan buildings, roof openings are essential for ventilation of spaces at the centre of the plan.

Planting

Planted roofs consist of three main components: a structural support including a water-proofing layer; soil; and a plant canopy (Fig. 1.30). Designing a successful planted roof requires an understanding of the interactions between these components and between the roof and the ambient environment. For the soil–plant system to have maximum effect on the building interior, the structural support should have little thermal inertia and be closely coupled with the soil. This implies that the soil–plant system should be capable of controlling heat transfer through the roof in a totally satisfactory manner. Where climatic conditions do not allow this, for instance where winters are cold, bulk thermal insulation is required. The introduction of thermal insulation decouples the soil–plant system from indoor spaces, consequently its effect indoors becomes indirect whilst it may continue to affect the microclimate outside.

Soil is a porous medium with a solid matrix (minerals and organic material), a liquid phase (water) and a gaseous phase (air and water vapour). In unsaturated soils heat will be transferred by conduction in the solid and liquid phases, by convection in the liquid and gas phases and by vapour diffusion in pores. Moisture content and temperature affect heat transfer of both the soil and the adjacent air layers. The plant canopy comprises the leaves and the air between and beneath them. A number of complex processes affect the thermal conditions in the canopy. These are discussed in Chapter 5 and a mathematical model is presented in Appendix A. Decisions that need to be taken during the design process include:

- soil type and depth, on the basis of its capacity to dampen temperature swings on a daily cycle (as a function of its thermal capacity, thermal diffusivity and thermal conductivity); its weight (for structural design considerations); and its suitability for the desired plants;
- selection of plants, on the basis of leaf density, leaf geometry and water requirements on a seasonal and annual basis; root structure etc.;
- irrigation scheme (where required);
- method of drainage for rainwater runoff.

Fig. 1.30 Planted roof of student residential building at Hooke Park, Dorset, UK. (Architect: E. Cullinan)

References

Allard, F. (Ed.) (1998) *Natural Ventilation in Buildings. A Design Handbook.* James & James (Science Publishers) Limited, London, 356 pp.

Alvarez, S., Lopez de Asiain, J., Yannas, S. and De Oliveira Fernandes, F. (Eds) *Architecture and Urban Space.*

Alvarez, S., Rodriguez, E. and Molina, J.L. (1991) The Avenue of Europe at Expo '92: application of cool towers. In International Conference Proceedings of the 9th PLEA. Kluwer Academic Publishers, Dordrecht, pp. 195–201.

Clark, E. and Berdahl, P. (1980) Radiative cooling: resource and applications. In Miller, H. (Ed.) *Passive Cooling Handbook.* Florida Solar Energy Association, pp. 177–212.

Francis, E. and Ford, B. (1999) Recent developments in passive downdraught cooling, an architectural perspective. In *European Directory of Sustainable and Energy Efficient Building.* James & James (Science Publishers) Limited, London.

Guerra, J., Alvarez, S., Molina, J.L. and Velazquez, R. (1994) *Guia Basico para el Acondiciomiento Climatico de Espacios Abiertos.* University of Seville.

Guerra Macho, J.J., Cejudo Lopez, J.M., Molina Felix, J.L., Alvarez Dominguez, S. and Velazquez Vila, R. (1994) *Control Climatico en Espacios Abiertos: evaluacion del proyecto Expo '92.* Universidad de Sevilla, CIEMAT, Junta de Andalucia.

Martin, M. (1989) Radiative cooling. In Cook, J. (Ed.) *Passive Cooling.* The MIT Press, Cambridge MA, pp. 85–137.

Meteotest (2003) *Meteonorm Version 5.0.* Global meteorological database for applied climatology. Meteotest, Bern.

Pearlmutter, D. (1993) Roof geometry as a determinant of thermal behaviour: a comparative study of vaulted and flat surfaces in a hot-arid zone. *Architectural Science Review*, 37: 75–86.

Pearlmutter, D., Erell, E., Etzion, Y., Meir, I. and Di, H. (1996) Refining the use of evaporation in an experimental downdraft cool tower. *Energy and Buildings* 23: 191–197.

2 ROOF CONSTRUCTION PRACTICES

TRADITIONAL TECHNIQUES

Traditional architecture is characterised by a wide range of styles, morphological elements and construction techniques. Though climate was one of the primary factors that informed indigenous methods, it is difficult to ascertain the sources of every single structural or morphological element. Locally available materials have contributed to the evolution of types and shapes of roofs and construction techniques applied. Pitched roofs prevailed in regions with high precipitation; their geometry has varied as a function of building typology, for example between detached and terraced buildings. Flat roofs are quite common in coastal and southern regions.

In ancient Greece wood was the main construction material for both flat and pitched roofs. Cedar and pine provided the beams on large spans; the theatre of Herodes Atticus in Athens was covered in cedar, the Parthenon with pine. As wood became more scarce it was mostly used in the north of the country, where it was more readily available and where pitched roofs were prevalent. An air gap or attic was provided between the roof structure and the ceiling of the living spaces below. Ochre or red ceramic tiles were used on the islands and the mainland regions of the Peloponnese, Macedonia, Thrace, etc. (Fig. 2.1, bottom). Grey slate is common in the regions of Pilion and surrounding islands, in Metsovo and Zagori (Fig. 2.1, centre and top). Ceramic tiles are commonly applied with sheathing and may indicate a ventilated roof construction, however sheathing is not necessary for the support of slate tiles.

Materials used for thermal insulation varied with location. Seaweed was in common use on the islands and in coastal areas, leaves in inland areas, small tree branches with leaves and bushes and goat hair in mountain areas. Sometimes mixed with soil and clay, these insulating layers were usually applied to flat roof construction but also, to a lesser degree, to pitched and vaulted roofs. Cane, a cheap and readily available material, was commonly used as a layer above the structural beams and joists (Fig. 2.2), supporting the mixture of materials that provided the thermal insulation layer.

On the island of Thira (Santorini), volcanic soil (known as Thiraic earth) provided the primary construction material. Structural materials and techniques for the vaulted roof types prevalent on the

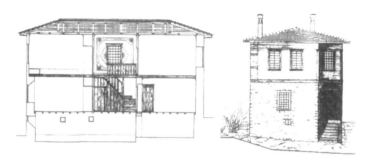

Fig. 2.1 **Typical traditional constructions of pitched roofs with timber structure and slate (top and centre) or ceramic tiles (bottom) covering from Northern Greece.**
(Source: Refs: Philippides C+P, 1984: Vol. 7)

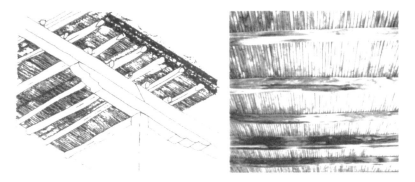

Fig. 2.2 Flat roof construction with logs, branches and canes.
(Source: Philippides, 1984: Vols 1 and 3)

Fig. 2.4 Views of the town of Oia, on the Greek island of Santorini showing vaulted and flat roof types.

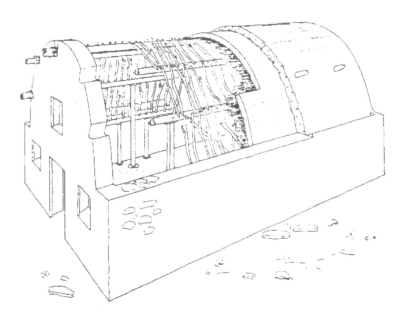

Fig. 2.3 Vaulted roof construction with Thiraic earth, seaweed, canes and wooden structural support.
(Source: Philippides, 1984: Vol. 2)

island are similar to those of the flat roofs shown in Figures 2.3–2.4. Typical construction for vaulted roofs includes the use of wood for the load-bearing structure of the canopy and the application of layers of branches and cane covered with a mixture of soil and dry seaweed for thermal insulation, finished (in the case of Santorini) with an external layer of Thiraic earth.

Fig. 2.5 Roofs in the mountain of the Pyrenees region of Spain. The ridges are oriented with the main wind direction.

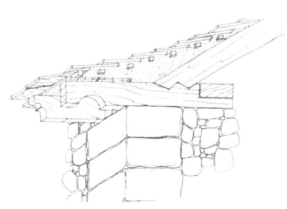

Fig. 2.6 Roof construction with local soil as an insulation layer.

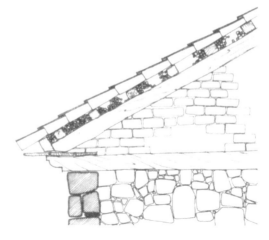

Fig. 2.7 Roof construction with local vegetal material for thermal insulation.

In Spain timber and stone were common in the north of the country and in the mountain regions. In southern and central regions, clay, soil and vegetal materials were plentiful, whilst wood was scarce and used for structural elements only. The geometry of pitched roofs took account of the main wind direction as well as precipitation (Fig. 2.5). Traditional roofs in mountain areas have large eaves with ridges facing in the direction of the prevailing winds. Where two or more main wind directions prevail, hip roofs are common, each with a different slope. Clay tiles are sometimes whitewashed to increase reflectivity. Lofts are commonly used for storing grain and food and are fitted with openings for ventilation. When the loft is used as a living area, the roof is provided with thermal insulation, using vegetal materials or compacted soil and gravel (Figs 2.6 and 2.7). The geometry and characteristics of the loft space have varied with region:

• in urban areas of the central region the loft is known as an *engolfa*, and is ventilated by south-facing openings under a ridge roof; in rural areas, it is known as *sobrado*, and its openings to the south are fitted with shutters to control ventilation and for protection from solar radiation;
• in the western mountain areas the loft, or *doblado*, has an air chamber which is vented through small openings under a ridge roof; it is also known as *sequero*;
• in the Pyrenees region, the loft has southern openings and a small window to the north. It is under a shed roof facing the main wind direction.

Unvented flat roofs are common in the southern and eastern coastal regions of Spain. Flat roofs are built with local materials, and their external finishing is usually of waterproofing soils. Thermal insulation is provided by layers of vegetal and mineral materials. In the southern arid zone and in the Baleares Islands, palm trees provide material for structural support, these are covered with leaves, seaweed (in coastal zones), small branches, *esparto* or vegetal waste (Fig. 2.8). Local soil and compacted clay are used as waterproofing materials. Annual maintenance, if required, is performed in conjunction with the whitewashing of buildings.

Fig. 2.8 Flat roof construction for arid zones: local waterproof soil as covering material, vegetal insulation layer and "pitacos" (arid plant trunk) structural support. Cabo de Gata, Almería.

Fig. 2.9 Typical Alpine roof construction with large overhang.

Fig. 2.10 The Malghe of the Valle d'Aosta.

In Italy, the traditional mountain refuges in the Pre-Alpine region are characterised by pitched timber roofs of shallow slope covered with wood flakes known as *scandole*. In the Alps area saddle roofs with slate tiles are mainly used with wide overhangs to prevent snow from settling on balconies and walls (Fig. 2.9). The loft is usually unheated; in winter, the snow that collects on the roof remains dry, thus acting as thermal insulation. In the Valle d'Aosta and other northern regions the seasonal dwellings, called *Malghe*, have roofs covered in large chips of stone (Fig. 2.10). These constructions, located at mid–high altitudes, are partially earth-bermed on the north side, utilising the difference in ground level and taking advantage of the thermal insulation provided by the ground. The layout of these constructions allows snow to slide directly onto the roof, thus providing greater protection from snowslides.

In regions with a mild climate (plains, coastal and hilly zones) pitched roofs are usually provided with wide overhangs. This is to protect walls and openings from precipitation, as well as to provide summer shading. A traditional roof typology in regions with hot dry climate (southern and inland regions) is flat stone construction disposed in concentric layers. Typical examples are the *Trulli* in Apulia (Fig. 2.11) and the *Nuraghi* of Sardinia (Fig. 2.12).

Fig. 2.11 Typical pointed covers in flat stone: the "Trulli" of Apulia.

Fig. 2.12 Archaic circular construction in stone, "Nuraghe" Sardinia.

CURRENT PRACTICE

Important characteristics of current practice are: the widescale adoption of concrete roof construction and the predominance of flat roofs (Fig. 2.13); the institutionalisation of thermal insulation as part of building regulations throughout the European Union and in many other countries; the rising awarenesss of environmental criteria in the selection of building materials and construction processes. Problems regarding moisture, condensation and thermal bridging have been contained or prevented. The enforcement of building regulations in support of energy conservation has increased the range of roof systems available. There are many constructional techniques that can combine thermal performance with load-bearing and waterproofing properties. Some typical examples from Greece and Italy are shown in Figs 2.14–2.16. The Spanish "*a la catalana*" roof is a vented roof that evolved from the traditional architecture of the region of Cataluña (Fig. 2.17). A set of weepholes within the parapet of the roof allow cross-ventilation through the air space. This can be adapted to current building codes by adding an insulation layer over the slab. There are increasing applications of planted (or "green") roofs which have proven popular in some parts of Europe (Fig. 2.18; see Chapter 5).

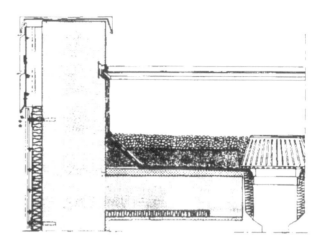

Fig. 2.14 Inverted flat roof from Greece.
From the bottom to the top: 1. reinforced concrete slab, 2. lightweight concrete, 3. water barrier, 4. thermal insulation, 5. gravel, 6. water protection layer, 7. protective wedge, 8. metal plate guide, 9–10. gutter

Fig. 2.13 Multi-storey apartment buildings of the 1950s and 1960s of reinforced concrete slab and beam construction with flat roofs.
(Source: Fatouros *et al*, 1984)

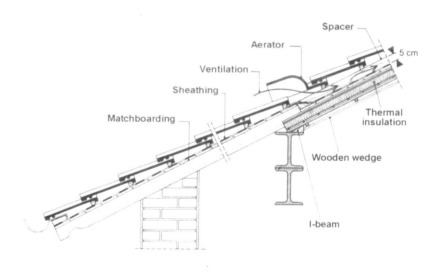

Fig. 2.15 Vented pitched roof from Italy.

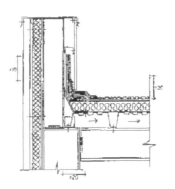

Vented pitched & flat roofs
Reinforced concrete slab
Vapour barrier
Thermal insulation
Protection cover with tar
Lightweight concrete
Thermal insulation
Vented air space

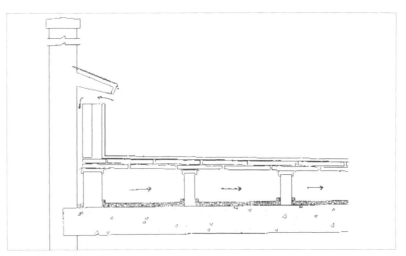

Fig. 2.17 Vented flat roof "a la catalana".

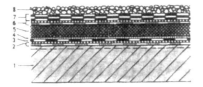

Compact flat roof
1. reinforced concrete
2–3–4. vapour barriers/waterproofing
5. thermal insulation
6–7. water protection
8. gravel

Lightweight flat roof
1. corrugated metal sheet
2. vapour barrier
3. tar layer
4. thermal insulation
5. dampening layer
6. water protection
7. gravel

Fig. 2.16 Pitched and flat roof variants from Greece.

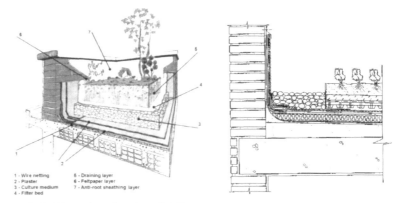

1 - Wire netting   5 - Draining layer
2 - Plaster   6 - Feltpaper layer
3 - Culture medium   7 - Anti-root sheathing layer
4 - Filter bed

Fig. 2.18 Planted roof variant details.

References

Colegio Oficial Arquitectos de Madrid. *Rehabilitación. La Cubierta*, Publicaciones COAM.

Fatouros, D.A., Papadopoulos, L. and Tentokali, V. (Eds) (1979) *Studies on the Dwelling in Greece*. Paratiritis, Thessaloniki.

Philippides, D. (Ed.) (1984) *Greek Traditional Architecture*. Vols 1–6. Melissa Publishers, Athens.

3 ROOF PONDS

INTRODUCTION

The term *roof pond* is commonly used to denote a system that incorporates a pool of water as a means of heat storage and as heat exchanger, or interim heat sink, for a building. Heat transfer between a roof pond and the occupied spaces in a building may be by means of direct *thermal* coupling or indirectly by the mediation of a separate element. In a cooling mode the water in the roof pond acts as a temporary repository of excess heat from the parent building, thus contributing to heat dissipation and cooling of indoor spaces. In turn, the roof pond water is cooled by natural processes.

Historically, in hot dry regions, the wetting of flat roofs by spraying or flooding has been a regular activity to induce direct evaporative cooling on hot summer days. Scientific investigation of roof ponds, with the water contained within plastic bags and cooling mainly by longwave radiation, began with the experiments and applications directed by Harold Hay since the 1960s (Hay and Yellott, 1969; Hay 1978, 1981a,b, 1985, 1986). Several patents were taken by Hay and others on a number of cooling and/or heating systems using roof ponds. From the mid-1970s a number of full-scale applications followed in the USA, mainly on newly built detached residential buildings; these were monitored and assessed (Marlatt *et al.,* 1984). As yet there are few recorded applications in Europe. The climatic applicability and performance of roof pond systems are reviewed in this chapter.

Appropriate choices of design parameters determining the system can allow occupied spaces to be maintained at stable and comfortable temperatures.

ROOF POND SYSTEM TYPES AND COMPONENTS

The main components of a roof pond system include the following, as shown in Fig. 3.1:

1 pond support;
2 water container;
3 protective cover;
4 spraying system and water recirculation.

Pond support
The building elements in contact with the roof pond should have a

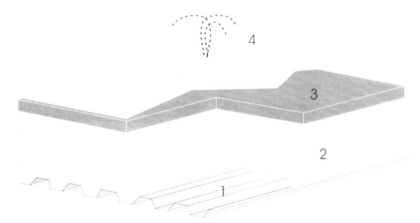

Fig. 3.1   Roof pond system components.
1 pond support
2 water container
3 pond cover
4 spraying system

high thermal conductivity to provide close thermal coupling between the water and the occupied spaces. This can be provided by a metallic ceiling or a reinforced concrete slab. Metals have higher thermal conductivity and can be used in much thinner layers, thus providing better thermal coupling without the additional thermal inertia given by a thicker element such as a concrete slab.

Water container
The water may be contained in plastic bags, in which case it will not be exposed to evaporative fluxes and will not require replenishment. Alternatively, water may be contained within the roof parapet over a watertight lining, its surface exposed to ambient conditions, in which case a water supply is required to replenish the water consumed by evaporation.

Protective cover
In the warm period of the year use of an insulated cover over a roof pond during daytime can protect the water from unwanted solar gains. The cover may be fixed or movable. An alternative is the use of floating insulation panels.

Spraying and water recirculation
Spraying circulates water from the roof pond, injecting droplets to the air above the pond; both the water and the adjacent air layers become cooler by evaporation.

ENVIRONMENTAL DESIGN PRINCIPLES

Cooling Mode

In the cooling mode the roof pond acts as an interim heat sink, with heat from occupied spaces rising naturally and being transferred to the water through the structural ceiling (Fig. 3.2, left). Water can store some 4166 kJ (1157 Wh) of heat per m^3 of its volume and degree rise in temperature, more than twice as much as can be stored in a slab of dense concrete (2000 kJ/m^{3}K or 555 Wh/m^{3}K). The water in the pond cools at night by longwave radiation to the sky (Fig. 3.2, right). During daytime it is essential to protect the pond from solar radiation and from high outdoor temperatures. This can be achieved with the use of a protective cover or other means of insulation. Spraying of the pond provides a further means of cooling, by evaporation.

Heating Mode

In climates with mild winters exposure of the pond to solar radiation on sunny winter days may warm the water sufficiently to contribute heat to the spaces below. At night, and at times of low solar radiation or low heat demand, use of the cover will insulate the pond from outdoor conditions, thus reducing heat losses from occupied spaces.

ADVANTAGES AND DISADVANTAGES

Advantages

- A roof pond system can be designed to provide both heating and cooling with the same components.
- Performance is independent of building orientation (this applies to all roof-based systems).
- Under favourable climatic conditions cooling performance is good and conducive to stable indoor temperatures within the comfort zone.

Disadvantages

- Lack of experience by the construction industry.
- The roof must support a load of 200–400 kg/m^2.

Other Design Considerations
On multi-storey buildings, roof pond systems can take cooling loads off intermediate floors by use of cooling panels.

APPLICATION TYPES

The known applications of roof ponds can be classified according to whether the water is contained in bags or within the roof parapet. Under each of these categories the thermal insulation may be fixed or movable. A US publication (Marlatt *et al.*, 1984) classified roof ponds into three main categories:

- *dry* when the water is contained in plastic bags;
- *wet* when the plastic bags are sprayed or flooded;
- *open* when the water is contained within the roof parapet.

The main applications are introduced below.

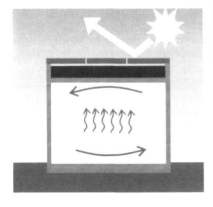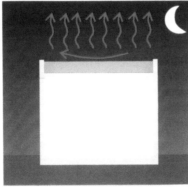

Fig. 3.2   Roof pond in cooling mode.
Left, daytime; right, nightime

ROOF POND EXAMPLES

dry
Skytherm House, Atascadero, California, USA
Bruder House Phoenix, Arizona, USA
Trinity University Test Facility, Texas, USA
Camelback School, Arizona, USA

wet
Universidad Pompeu Fabra, Barcelona, Spain

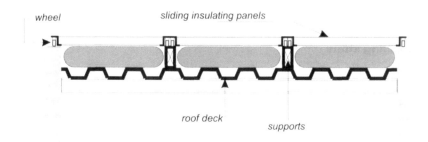

Fig. 3.3 Skytherm components.

WATER CONTAINED IN BAGS: **Skytherm**

* Hay, H. US Patent
3299589 (Jan 24,
1967).

The roof pond concept patented by Harold Hay as *Skytherm** in 1967 combined the traditional functions of a roof with an effective natural heating and cooling system. In a typical application a corrugated metal ceiling deck was used for structural support for the pond (Figs 3.3 and 3.4). The entire roof area of the building was treated with roof ponds. The water was contained in plastic bags with typical water depths of 100–300 mm. A watertight liner or coating was placed over the metal deck as additional protection from water leakage. The movable insulation panels were commonly of 50–75 mm rigid foam and covered with a protective skin. The panels were operated manually or by automatic control, and a mechanical guidance and drive system physically relocated the panels in either position.

The Skytherm system was applied successfully in different parts of the USA (see examples on the following pages). In the cooling mode, the insulation panels were closed during the daytime to protect the roof and building interior from solar radiation and from transmission heat gains (Fig. 3.4, left). The panels were retracted at night to allow the pond to cool by thermal radiation to the sky and by convection to the outdoor air (Fig. 3.4, right).

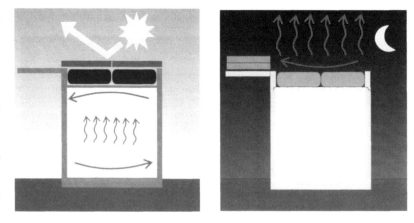

Fig. 3.4 Skytherm system summer cooling modes.

SKYTHERM HOUSE

Atascadero, California, USA

Figure 3.5 shows the prototype Skytherm system built in 1973 under the supervision of its inventor. The building was monitored and was shown to provide a very good performance, maintaining stable and comfortable temperatures throughout the year, and did not require conventional space heating or cooling. Roof ponds maintained internal space at average temperatures of 22°C (72F) in summer (Fig. 3.6) and 19–23°C (66–74F) in winter. Occupants considered that conditions achieved were superior to those provided by conventional air conditioning (Marlatt et al., 1984).

Fig. 3.5 Skytherm House, Atascadero; photograph soon after completion showing roof ponds (left) and the house today (right).

Client: H. Hay
Architects: K. Haggard, J. Edminsten, P. Niles
Date completed:1973
Floor area: 185 m²
Roof pond area: 102 m²

Key Features

- Four 215 mm (8.5 inch) deep waterbags, each 2.44 m × 11.60 m.
- Top layer of transparent PVC above waterbags.
- Movable, custom-made, insulating panels operated by an automatic differential thermostat control system, developed specifically for the building.

Roof Construction

- Corrugated metal deck.
- Waterproofing layer: fibreglass reinforced asphalt.
- Waterbag material: PVC.
- Panel construction: acrylic covered polyurethane in steel framework.
- Guidance system: stacking.
- Drive system: motor, sprocket, chain.
- Control system: differential controller.

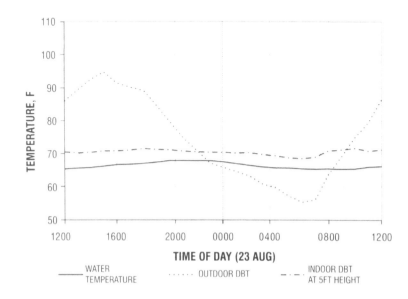

Fig. 3.6 Measured temperatures showing performance of the Skytherm system.

BRUDER HOUSE
Phoenix, Arizona, USA (Fig. 3.7)

The design of this building was developed directly from the proto-type house at Atascadero.

Key Features

- Prototype use of panel stacking system, galvanised steel track beams, lightweight drop-in panels, multi-zoned pond areas, and cable drive system.
- Roof pond area is equal to floor area of house.
- Ceiling consists of standard metal decking.
- The insulating panels move on top of one another when rolled back into stacked position.
- Track system designed to prevent lifting-off of pond panels in high winds.
- A double-belt drive system moves the insulating panels on roof tracks of galvanised metal.
- The roof is lined to protect against water leakage.
- Manual spraying using roof-mounted sprinklers.

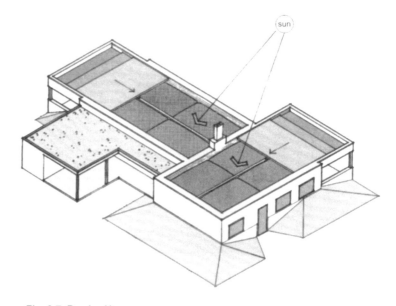

Fig. 3.7 Bruder House

Client: E. Bruder
Architect: D. Aiello,
Janus Associates
Date completed:1979
Floor area: 149 m²
Roof pond area: 149 m²

Roof Construction

- Corrugated metal deck.
- Waterproofing layer: polyethylene.
- Waterbag material: clear PVC.
- Panel construction: aluminium foil covered Thermax 604 in galvanised steel framework.
- Guidance system: stacking.
- Drive system: motor/belt/cable.
- Control system: differential controller.

TRINITY UNIVERSITY TEST FACILITY
San Antonio, Texas, USA (Fig. 3.8)

Key Features

- Ten water bags each 6.10 m × 1.22 m, are supported by a thin, deeply corrugated steel deck, the underside of which forms the ceiling.
- The custom made water bags are of UV-stabilised PVC; three layers of plastic used for the bags, the top layer forms an air cell on the bag.
- A 45-μm (1.14 mm) black rubber liner is inserted between the water bags and the deck.
- Prefabricated 2.44 m × 3.05 m sections are bolted together to form the two 3.05 m × 12.2 m movable insulation panels which move along galvanised steel tracks attached to the parapet edges, one riding higher and overlapping the other, to and from a stacking area adjacent to the roof.
- The insulating panels are composed of corrugated galvanised steel bonded on a rigid insulating board with a layer of foil on the underside of the board.
- A motor-operated, tooth and gear drive draws the panels across the sloping roof.

Roof Construction

- Corrugated metal deck.
- Waterproofing layer: EPDM.
- Waterbag material: clear PVC.
- Glazing material: PVC air cell.
- Panel construction: corrugated steel in steel framework.
- Guidance system: stacking.
- Drive system: motor/rack/pinion.
- Control system: manual.

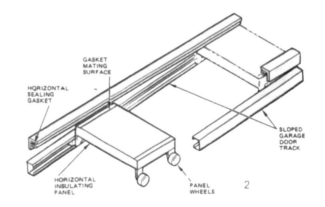

1

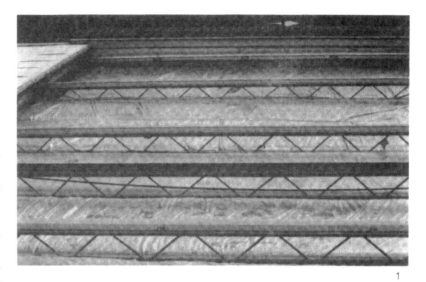

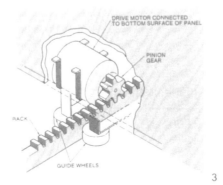

2

3

Fig. 3.8 Trinity University Facility

1 Waterbags;
2 detail of sliding insulation system;
3 rack and pinion system.

Client: Trinity University
Architect: D. Bentley
Date completed: 1980
Floor area: 74 m²
Roof pond area: 74 m²

CAMELBACK SCHOOL
Paradise Valley, Arizona, USA (Fig. 3.9)

Key Features

- The roof pond comprises ten 254-mm-deep waterbags placed on a flat metal sheet welded to the corrugated metal roof deck.
- The roof bays are lined and waterproofed.
- Steel "I" beams support the decking and the roof pond system with the decking resting on the lower flange of the beam.
- Three insulating panels, each 7.32 m wide, span the entire width of the building.
- A spray system to wet bags in summer for evaporative cooling.
- A cable drum and motor drive system controlled by a timer operate the panels. The cable system moves under the panels rather than above them.

Fig. 3.9 Camelback School.

1 Detail section;
2 roof pond and insulating panels.

Client: J. Rosenheim
Architect: D. Aiello
Date completed:1982
Floor area: 35 m²
Roof pond area: 35 m²

Roof Construction

- Corrugated metal deck.
- Waterproofing layer: asphalt extended urethane.
- Waterbag material: clear PVC.
- Panel construction: foil-covered Thermax 604 encased in a galvanised steel framework.
- Guidance system: stacking.
- Drive system : motor/rack/pinion.
- Control system: manual.

Water Pond Contained Within Parapet

Figure 3.10 illustrates the installation of a roof pond tray on an experimental test rig in Almeria, Spain.

Fig. 3.10 Experimental roof pond in Almeria, Spain.

The tray is being carried to the test rig (top left), waterproofed (top right), filled with water (above left) and the spraying system is set in operation (above right).

This experimental installation was part of the ROOFSOL project that was undertaken in the context of the European Commission's Joule research programme.

ENERGY ROOF

A system patented by A.L. Pittinger and W.R. White in the USA in the 1970s with a roof pond supported on a metal ceiling and thermal insulation floating on the water under a thin, transparent plastic film (Fig. 3.11; Marlatt *et al.*, 1984).

For summer cooling, at night water is pumped to a distributing tube, which allows it to flow in a thin layer above the insulation where it is then cooled by longwave radiation (Fig 3.11 b/c). A water spray promotes evaporative cooling at the upper surface of the plastic film. In winter the pump is operated only during daytime for heating.

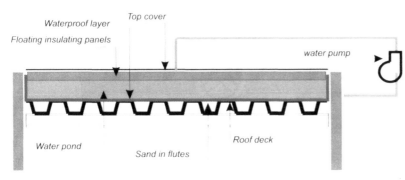

(a) Cross-section

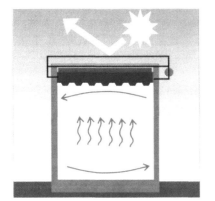

(b) Summer day operation

(c) Energy roof cooling

Fig. 3.11 Energy roof.

UNIVERSIDAD POMPEU FABRA
Barcelona, Spain

Key Features

- The water pond is contained within the roof parapet above a pre-existing roof basin.
- The roof pond is filled with rain water.
- A water propulsion system is fitted at the bottom of the roof pond.
- The roof basin is filled with approximately 400 mm of water.

Roof Construction

- Structural support for roof pond on top of old massive brick basin.
- Air space between old structure and bottom of roof pond.
- Perimeter concrete slabs 50 mm thick form the base of the roof pond.
- Waterproofing layer: heavy duty PVC.
- Surface treatment: laminated stainless steel.
- Structural support of galvanised steel for perimeter installations.
- PVC concealed gutter for water overflow.

CIC,Construction Monthly N° 302, April 1997.

Client: Universidad Pompeu Fabra
Architect: Clotet Paricio Arquitectos
Date completed: 1995
Floor area: 1225 m²
Roof pond area: 840 m²

Fig. 3.12
Universidad Pompeu Roof pond.

EXPERIMENTAL SYSTEMS

An experimental system (Erell and Etzion 1998a) tested at Sde-Boqer, Israel consisted of a 200-mm-deep roof pond on a 120 mm reinforced concrete roof (Figs 3.13–3.14). The water was contained in a waterproof PVC membrane lining, and was covered by floating polystyrene insulation 100 mm thick to reduce daytime heat gain from solar radiation and ambient air. Small gaps between the polystyrene boards allowed water sprayed above the insulation to trickle into the pond. A series of white PVC panels supported on metal frames served as shading devices and wind deflectors. The panels were sized so that each row was larger than the one in front on the windward side, channelling the wind past the rows of sprayers mounted on the same frames at a height of 50–100 mm above the water. Thus, each row of sprayers was exposed to essentially fresh, dry, ambient air.

The system combined features of both direct and indirect systems. In a direct mode, an opening at the centre of the roof allowed the intake of cool, moist air directly into the building. Air was sucked in

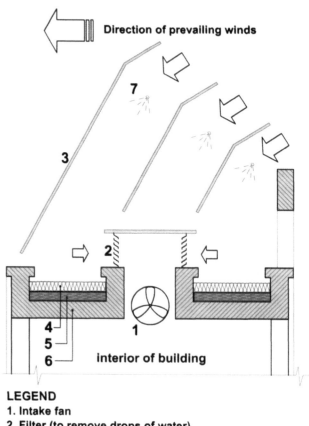

Direction of prevailing winds

LEGEND
1. Intake fan
2. Filter (to remove drops of water)
3. Wind deflector (and sunshade)
4. Floating polystyrene
5. Water in roof pond
6. Concrete roof
7. Water sprayer

Fig. 3.14 Schematic section of the system tested.

Fig. 3.13 Evaporative cooling roof prototype at Sde-Boqer.

by means of a small, low-powered electric fan, after being filtered by a coarse mesh to remove suspended drops of water (Fig. 3.14). Evaporated water was replaced from the mains supply. The depth of water in the roof pond was controlled by means of a float valve. Water was filtered to remove dust and foreign objects that might clog the pumps. In the indirect mode, the system could be used to cool water which had absorbed heat from the room below conducted through the concrete roof. The water in the roof pond could be used also to cool non-adjacent interior spaces by means of water–air heat exchangers. The operating schedule of such a system depends on whether it is designed mainly to cool water or air.

The system's cooling effect was assessed with and without spraying. Sprayers were initially operated continuously for several days. In this mode a clear daily pattern was observed with the roof pond following the ambient wet-bulb temperature at an offset of between 3°C and 4°C (Fig. 3.15). The water in the pond cooled as the wet-bulb temperature fell. However, continued operation of the sprayers during daytime resulted in a gradual increase in pond temperature in line with the change in the wet-bulb temperature. When the sprayers operated only at night (2000–0500), it was possible to keep the pond water closer to the ambient wet-bulb temperature, some 3°C cooler than when the sprayers operated continuously (Fig. 3.16). Thus for the purpose of cooling the pond water, night operation of the sprayers was the preferred mode. This reduced the amount of water evaporated, used less electricity for pump operation and was more effective in cooling the roof pond. Daytime spraying would be recommended for times when the water temperature in the roof pond had risen well above the ambient wet-bulb temperature. The results of this experiment suggest that a difference greater than 4°C between pond and ambient wet-bulb temperatures would justify continuing sprayer operation during the daytime.

The amount of water evaporated could not be measured accurately but was estimated at 30–50 litres per day. The difference between the air temperature in the test room and a reference room of similar construction was the result of a heat flow rate of approximately 10–15 W/m² through the test room ceiling, or about 100–150 W (Fig. 3.17). The total daily cooling output was in the range of 2.5–4.0 kWh. In the process of cooling water, the evaporative system also produced cool, moist air. In the last stage of the experiment this air was drawn directly into the test room. The effect was to lower the room temperature to within 1–2°C of that of the roof pond water, or about 1°C lower than was possible by cooling the ceiling alone. Compared with the reference room, the mean daily temperature was about 4°C lower (Fig. 3.18).

The system produced greater cooling output when the initial operating temperature of the roof pond was higher. The difference between the roof pond temperature and the ambient wet-bulb temperature was approximately 4°C on average, and this limited the cooling output of the system. Performance may be improved by different types or combinations of roof sprayers.

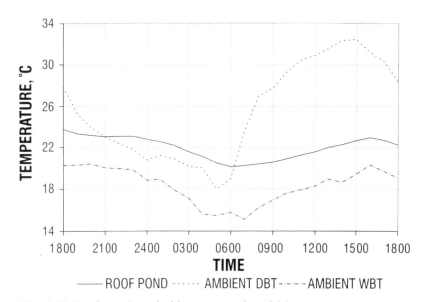

Fig. 3.15 Roof pond cooled by evaporation, 24-hour sprayer operation.

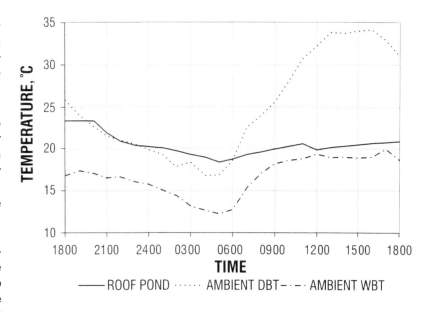

Fig. 3.16 Roof pond cooled by evaporation, operating 2000–0500.

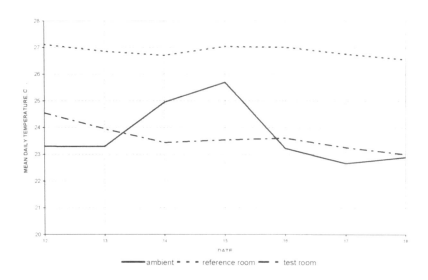

Fig. 3.17 Effect of indirect evaporative cooling system on room temperature over a period of six days in August 1997.
Sprayers operating 2000–0500

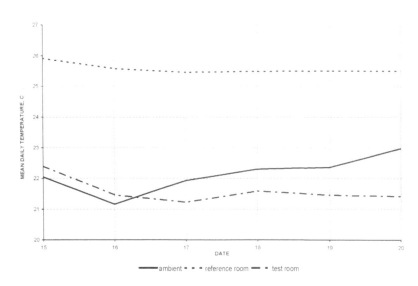

Fig. 3.18 Effect of evaporative cooling with forced ventilation on room temperature over a period of five days in September 1997.
Cooling system operative 2000–0400 sprayers and forced ventilation 2000–0400

DESIGN CONSIDERATIONS

A detailed mathematical model was developed to investigate the performance of roof ponds with different design parameters and operational conditions (Rodriguez et al., 1998). The model is included in the software described in Appendix B. An abridged version of the model is presented in Appendix A. Simulation studies performed with the model provided the basis for the design guidelines and performance data given in this section and in Chapter 6. The design parameters considered were the following:

- water depth and solar absorptivity of the pond floor;
- pond support structure;
- physical properties and operation of pond cover;
- spraying system and water recirculation.

These are discussed below and design guidelines are summarised in Table 3.1. Numerical results illustrating the likely performance of different roof pond variants on single-storey and multi-storey residential and commercial buildings in different locations are presented in Chapter 6.

Water Pond Depth and Absorptivity

Pond depth determines the volume of water that can be contained per unit roof area, and thus the heat storage capacity and thermal inertia of the system. Depth and volume of water are also a measure of the structural loads imposed by the pond. Common values in practice have been in the range of 250–500 mm, with the higher figure more appropriate for uncovered ponds and the lower for protected ponds. The simulations considered pond depth in the range 100–700 mm. It was found that for ponds that are shaded by a protective cover during daytime, and/or those cooled by sprays (see also below), pond depth has a lesser effect on performance. In such cases, depth may be determined by water availability or by structural considerations. Unshaded ponds are exposed to higher heat gains and thus require additional water to prevent overheating. The colour of the pond floor affects the solar absorptivity of this surface. Light colours are preferred, to reflect radiation, especially in the case of uncovered ponds. However, the sensitivity analysis showed that most of the incoming radiation is absorbed by the water before it reaches the pond floor. Thus this parameter has a negligible effect, especially in the case of shaded ponds.

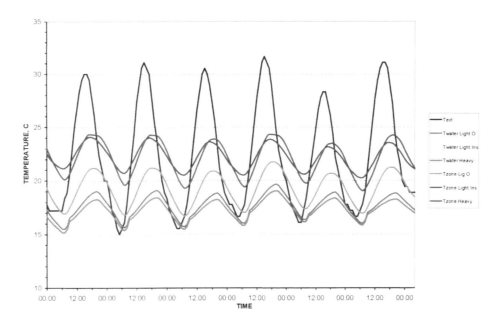

Fig. 3.19 Pond water temperature and indoor zone temperatures as a function of different roof types over a period of six summer days in Seville.

Pond Support Structure

The simulations showed that the higher the conductance of the supporting structure, the higher the cooling effect at night. Hence this should be a thin layer of material of high thermal conductivity. In practice, however, the pond structure might consist of a thicker layer such as a concrete slab, as in the experimental example described above, and may also include a layer of thermal insulation to reduce winter heat loss. Figure 3.19 illustrates the effects of these parameters on indoor temperature and on the pond water temperature. Three different types of roof structure were considered in conjunction with a roof pond of 0.30 m depth: an insulated heavyweight roof structure (indicated as "Heavy" on the graph), an insulated lightweight structure ("Light Ins") and an uninsulated structure closely coupled to the pond ("Lig O"). Detailed descriptions of these are given in Chapter 6. In all three cases an insulated cover of 0.100 m thickness and 0.037 W/mK thermal conductivity is assumed to be drawn over the pond during daytime and no spraying is in operation at any time. Whilst the outdoor air temperature fluctuated from 15°C to over 30°C the water pond temperature has much less fluctuation and keeps quite cool in the range of 15–20°C. All three roof structures are shown to provide comfortable indoor temperatures, the well-coupled uninsulated ceiling has the coolest.

Pond Cover

The sensitivity studies investigated the thermal insulation properties of the cover with values of thermal conductance in the range 0.02 W/m²K (extremely high insulation) to 20 W/m²K (no insulation at all). Solar absorptivity was kept constant at 0.3 to emulate a light-coloured surface exposed to outdoors and the solar transmissivity was varied between 0.0 and 0.6. The radiative transfers were considered as a function of the upper and lower emissivities of the cover surfaces which were varied in the range of 0.3 (low emissivity) to 0.9 (typical of building surfaces). An opaque cover provides better protection from solar radiation. Where total opacity is not possible, a material with a solar transmissivity no higher than 0.2 should be selected. Moreover, unless the pond cover contains insulating materials, its external surface should be of a light colour to reflect solar radiation. The following are the main considerations relating to the operation of the cover:

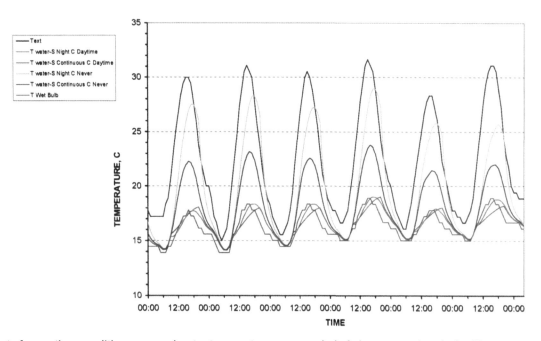

Fig. 3.20 Effect of operating conditions on pond water temperature over a period of six summer days in Seville.

1. *Movable cover:* opening the cover to expose the pond at night and closing it during daytime can provide a good performance without a spraying system.
2. *Fixed cover:* with a cover permanently in place and separated from the water level by a ventilated air layer the radiant cooling potential is lost, but the convective and evaporative effects apply; in some locations these could be sufficient to remove all the thermal loads without need of a spraying system.
3. *Fixed floating insulation:* this is cheap and easy to apply, and the thickness of the insulation can be specified as needed to minimise the heating effect of solar radiation; the insulation does, however, inhibit heat dissipation at night and a spraying system becomes essential.

Spraying System

The design and operational characteristics of the spraying system have a strong influence on the performance of the system. These are summarised below.

1. *Droplet radius:* very small droplets tend to cool to the ambient wet-bulb temperature, resulting in excessive water consumption; on the other hand, very large droplets may not produce a sufficient cooling effect. The range of 0.5–1.0 mm was found to be adequate. The suppliers of the spraying nozzles should guarantee the radius produced by their products.
2. *The height of spray:* spray heights in the range 0.5–1.0 m are sufficient to allow the droplets to cool as they fall.
3. *Horizontal distribution of nozzles:* nozzles should be distributed uniformly around the pond surface; the arrangement must avoid overlap to conserve energy and reduce the waste of water.
4. *Pond water recirculation:* for a pond that is coupled to a single space the studies suggested that the volume of water in the pond should be circulated once per hour with higher rates for increased loads.
5. *Operational strategy:* the spraying system should be turned off when the pond temperature falls below the wet-bulb temperature since in such a case spraying would lead to a warming of the pond.

Operational Conditions and Roof Pond Variants

The effects of using a pond cover and spraying system are illustrated in Fig. 3.20. The cover properties are as for the cases shown in Fig. 3.19. The spraying system is characterised by a drop radius of 1.0 mm, spray height of 1.30 m and one recirculation of the pond water per hour. The graph summarises the results of hourly simulations for the same mid-summer week in Seville. When the roof pond is left uncovered spraying has a significant cooling effect, especially when continuous. However, using an insulated cover during the daytime has a much more substantial effect, reducing fluctuations as well as peak temperatures. In fact, the simulations suggest that if there is a cover there is little point in using a spraying system, especially during daytime.

As spraying systems consume water as well as electric power, it is desirable to limit the duration of their operation. On roof pond systems that are exposed to the sky at night and are thus also cooled by radiation, spraying should be stopped when the pond temperature reaches the ambient wet-bulb temperature. On roof pond systems that are not exposed to the sky at night (for example, systems with floating insulation) spraying should be stopped when the water temperature is 3–4°C above ambient wet-bulb.

Table 3.1 summarises design considerations for roof pond systems.

References

Erell, E. and Etzion, Y. (1998a) *Development and Testing of an Evaporative Cooling Prototype*. ROOFSOL Roof Solutions for Natural Cooling. Final Report. European Commission Joule Programme.

Hay, H. (1978) A passive heating and cooling system from concept to commercialisation. *Proceedings of the Annual Meeting of the American Section of the International Solar Energy Society*.

Hay, H. (1981a) Integration of radiative systems. *Passive and Hybrid Cooling Notebook*. Florida Solar Energy Association, USA.

Hay, H. (1981b) Thermopond: applicability to climate and structure. *Passive and Hybrid Cooling Notebook*. Florida Solar Energy Association, USA.

Hay, H. (1985) Roof ponds ten years later. *Proceedings of the 10th National Passive Solar Conference*. American Solar Energy Society, USA, pp. 181–186.

Hay, H. (1986) Roof ponds: the humidity issue. *Proceedings of the 11th National Passive Solar Conference*. American Solar Energy Society, USA, pp.181–186.

Hay, H. and Yellott, J. (1969) Natural air conditioning with roof pond and movable insulation. *ASHRAE Transactions*, pp. 165–177.

Marlatt, W., Murray, K. and Squier, S. (1984b) *Roof Pond Systems*. Energy Technology Centre, California, USA.

Rodriguez, E., Molina, J.L., Guerra, J.J. and Esteban, C.J. (1998) *Detailed Modelling of Water Ponds*. ROOFSOL Project Task 2 Final Report. European Commission Joule Programme.

TABLE 3.1 **SUMMARY DESIGN CONSIDERATIONS FOR ROOF PONDS***

UNCOVERED & NO SPRAYS	UNCOVERED & SPRAYS	COVER & NO SPRAYS	COVER & SPRAYS
The water pond is permanently exposed to ambient air without a cover, no spraying system.	*Spraying system operating day and night over uncovered pond to provide a cooling effect.*	*Pond covered during daytime; no spraying.*	*Pond covered during daytime with spraying operated at night only.*
Parameters assessed: • reflectance of pond floor • water depth	Parameters assessed: • droplet radius • water flow rate • maximum height of spray	Parameters assessed : • thermal conductance of cover • emissivity of cover • solar absorptance of cover • solar transmissivity of cover • air space between cover and pond	Parameters assessed: • droplet radius • water flow rate • maximum height of spray
Recommendations • Water depth should be at least 300 mm • Water temperature will increase because of the solar gains until compensated by spontaneous evaporative effect; typical water temperature fluctuation is around 10K.	**Recommendations** • Droplet radius in the range 0.5–1.0mm should be adequate • Water flow rate for sprays should be 1.0–1.5 volumes per hour • The minimum height of the spray should be 0.5 m • Limiting spray operation to night-time can conserve water, but is required to maintain a stable water temperature in shallow ponds (<300 mm). For deeper ponds the increase in water temperature during daytime will be less than 7–8K, even in warm and sunny conditions.	**Recommendations** • Pond cover prevents over-heating of water, whilst spontaneous evaporation lowers water temperature below ambient average • The emissivities of the cover surfaces have negligible effect • For opaque covers solar absorptance of cover has little effect on water temperatures and cooling • Ventilation of the airspace between cover and water surface does not affect performance.	**Recommendations** • Droplet radius 0.5–1.0mm with water flow rate 1.0–1.5 pond volumes per hour • Pond cover reduces fluctuation in pond temperature whilst spraying lowers pond temperature at night; cooling rate is higher compared with uncovered pond with sprays. • Spraying should be stopped during daytime unless water temperature increases substantially above the ambient wet-bulb.

* See Chapter 6 for performance data and applicability maps.

4   COOLING RADIATORS

INTRODUCTION

In the context of passive cooling, the term radiator denotes a device that is designed to dissipate heat from a building by means of long-wave radiation to the sky. The radiator may be an integral part of the building, or a device designed especially to dissipate heat from a heat transfer fluid, such as air or water. To cool a building, the radiator must be in good thermal contact with indoor spaces, or act in conjunction with an element of high thermal capacity that provides interim heat storage.

RADIATOR SYSTEMS AND COMPONENTS

The main components of a cooling radiator system include the following (Fig. 4.1):

1. Radiator (heat emitter or dissipator);
2. Heat transfer fluid (water or air);
3. Thermal insulation;
4. Cool store;
5. Heat exchangers with the building.

The thermal processes are illustrated in Figure 4.2.

Emitter

The emitter may form part of the external finishing of a roof or it may be added as an independent component. To act as a heat dissipator, the emitting surface must have a high longwave emissivity. Typical painted surfaces, as well as most building materials (though not polished metals) have an emissivity of approximately 0.9. A reflective external finishing or use of an insulated cover are essential to reduce the impact of incident solar radiation during daytime. Aluminium, stainless steel, copper or lead have a high thermal conductivity, which is essential for good thermal coupling with the working fluid.

Heat Transfer Fluid

Heat transfer between the radiator and the fluid depends on the area of contact, the temperature difference, the properties of the fluid and the physical dimensions of the system. The specific heat of air is 0.28 Wh/kg °C (1.006 kJ/kgK) compared with 1.163 Wh/kg °C (4.183 kJ/kgK) for water, and its density (at a temperature of

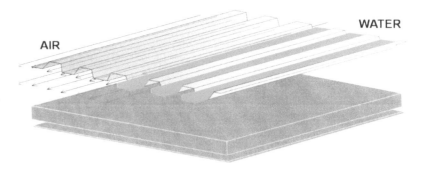

Fig. 4.1 Cooling radiator roof components: emitter, heat transfer fluid and thermal insulation.

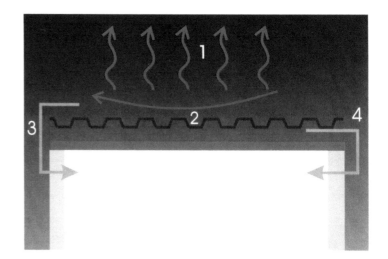

Fig. 4.2 Thermal processes

1 Net radiative transfer to sky (difference between incoming and outgoing longwave radiation)
2 Convective exchange between ambient air and radiator surface
3/4 Convective exchange between the radiator and fluids used for heat transfer

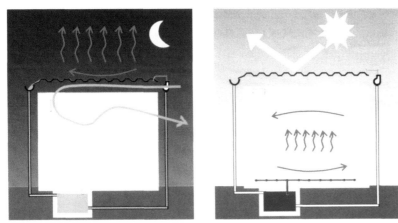

Fig. 4.3 Radiator in cooling mode: night (left) and daytime (right) operation.

300K or 27°C) is only 1.18 kg/mΔ compared with 996 kg/mΔ for water, which makes the latter a far more efficient heat transfer fluid. For this reason, air requires a far greater contact area for heat transfer than does water. On the other hand, air does not freeze and air leakage is far less critical than a water leak.

Thermal Insulation

Thermal insulation is generally required on the roof of the building to control undesirable heat loss in winter and to limit heat gains from the outdoor environment in summer. When a cooling radiator is installed on the roof, thermal insulation for the building may be applied below the radiator, in which case the radiator cools a heat transfer fluid and has no direct contact with indoor spaces. Alternatively, movable insulation may be installed above the radiator and removed when it is desirable to cool the roof.

Cool Store

The building structure, or a specially designed store, may be provided to allow "coolth" produced at night to be accessed during the day. With water radiators, an insulated water tank can be specified for this purpose. Air radiators can be coupled to a rock-bed storage medium.

Heat Exchangers

When the radiator is not a part of the roof itself, or when it is desirable to absorb heat from other parts of the building, heat exchangers must be added, through which a suitable heat transfer fluid is circulated. Cooling panels are one example of such an arrangement. These may be made of concrete, with embedded pipes or ducts for the circulation of water or air, and attached to intermediate floor slabs of multi-storey buildings collecting heat from a zone of a building. The heat is then transferred to a water or air radiator for dissipation at night.

ENVIRONMENTAL DESIGN PRINCIPLES

Cooling Mode

Cooling of the heat transfer fluid is obtained by exposure of the radiator to the night sky. The heat is dissipated by longwave radiation and, when the radiator surface is warmer than the ambient air, also by convection. During daytime, heat transfer between the radiator and the occupied spaces below the roof must be inhibited. Where daytime cooling is required it should be provided from a store charged from the "coolth" produced at night (Fig. 4.3). The effectiveness of a cooling radiator depends on the coupling of the building with the emitter. It also depends on environmental conditions, the ambient air temperature and the sky temperature, which in turn depend on atmospheric humidity and cloud cover.

Heating Mode

A radiator could also contribute to winter heating by acting as a solar collector, but the heavy radiative and convective heat losses it could incur may undermine the effectiveness of this operation.

ADVANTAGES AND DISADVANTAGES

Advantages
- Use of conventional components;
- Can be retrofitted over existing flat roofs at low cost.

Disadvantages
- Must be laid flat or slightly pitched;
- May require provision of a specially designed cool store;
- On buildings of more than one storey radiators need to be coupled to cooling panels acting as interim heat stores.

APPLICATION TYPES AND EXPERIMENTAL WORK

Trickle Roof

This system, devised by Harry Thomason in the USA, makes use of both radiative and evaporative cooling processes (Figs 4.4 and 4.5). A film of water is made to flow across a pitched, lightweight surface. The water is cooled by contact with the high-emissivity surface and by evaporation. The cooler water is then pumped into an insulated storage tank and during daytime circulated in the building through pipes in the floor or ceiling slab.

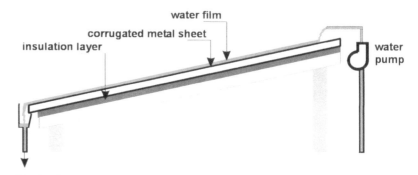

Fig. 4.5 Cross-section of trickle roof.

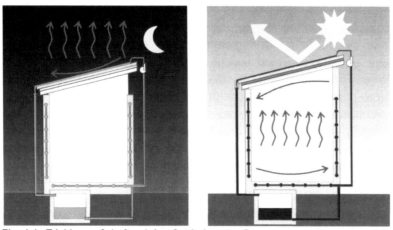

Fig. 4.4 Trickle roof. Left: night. Cooled water flows across the roof and is stored in tank; Right: stored cool water is circulated to building spaces.

Air-based Radiators

This roof system, shown in Figs 4.6 and 4.7, proposed by Givoni (Givoni, 1977) as a roof radiation trap, consists of a fixed, lightweight metal radiator, with air as the heat exchange medium. The system draws in ambient air at night, cooling the air as it comes in contact with the radiator. The cool air is then drawn inside the building by means of a small fan and is vented through exhaust outlets. The radiator roof, sloped to the north, forms a south-facing space with glazing and operable insulating shutter. On sunny winter days, this shutter is opened so that the roof space can function as a solar collector. On winter nights and during summer, the shutter is kept closed. One drawback to this system is that the thermal coupling

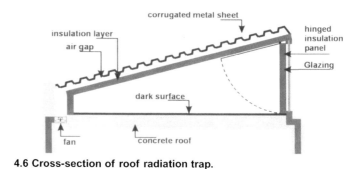

4.6 Cross-section of roof radiation trap.

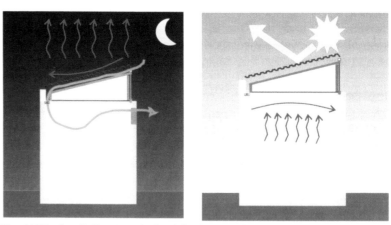

Fig. 4.7 Roof radiation trap. Left: night-time cooling mode; Right: daytime operation.

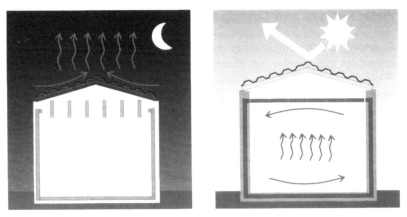

Fig. 4.8 **Air-based radiator. Left: night: insulation panels open for cooling by longwave radiation; Right: daytime: reflective metal and thermal insulation protect from sun.**

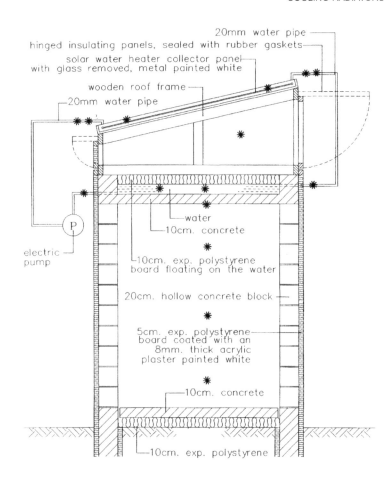

Fig. 4.9 Insulated roof pond with unglazed radiators.

between the radiator and the building's thermal mass is poor. A second drawback is that because air has a very low heat capacity its use as a heat exchange medium requires large volumes of air to be transported. The air channels and electric fans required for such a system impose fairly complex demands on the design of the building, as well as adding to its cost.

Figure 4.8 shows another system proposed by Givoni (1994), which uses movable insulation on the interior of the building. The system was designed for low-mass roofs, such as corrugated iron or clay tiles, as a low-cost solution. The insulating panels are hinged, and may be rotated from the interior. During daytime, the panels are in a horizontal position, reducing solar heat gains. At night, the panels are rotated to the vertical position, so that the roof is exposed to the building interior. After sunset, the low-mass roof cools down quickly, and the building interior is cooled by radiation and convection. The attractions of this system derive from the low cost and ease of installation of internal insulation panels compared with external insulation. The operation of the movable insulation system is quite simple; the system requires no additional changes to the building, structural or otherwise. However, coupling of the thermal mass of the building with the radiator is poor, so cooling rates may be expected to be modest. Regrettably, this system has not been tested.

Water-based Radiator With Roof Pond

A passive cooling system integrating radiative and convective cooling was tested at Sde-Boqer, Israel (latitude 30°48'N, longitude 35°58'E, altitude 475 m). The system was based on circulating water at night from a shallow, insulated roof pond 100–150 mm deep through radiators fixed above it (Fig. 4.9), by means of a small electric pump (Etzion and Erell, 1989; Erell and Etzion, 1992). The water was cooled and excess heat was removed from the building by longwave radiation and by convection. The water flowing through the system kept the radiators relatively warm, thus

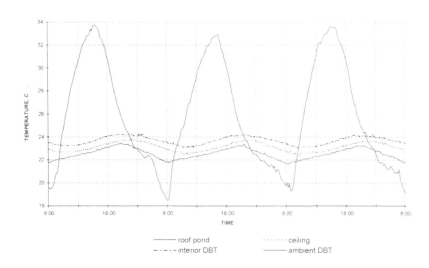

Fig. 4.10 **Temperatures of roof pond water, test cell ceiling and interior over a three-day experiment.**

increasing radiative losses and eliminating convective gains. The water in the roof pond served as part of the thermal storage mass of the building. Changing the amount of water in the system was thus a means of controlling the thermal mass in the building, and therefore its diurnal temperature swing. In addition to 100-mm-thick polystyrene panels floating on the surface, water in the roof pond was protected from solar radiation by the radiator panels, which doubled as fixed shading devices over the roof. The effect of radiative cooling on the temperature inside the test cell and that of the roof pond water is shown in Fig. 4.10. A winter heating mode was also tested for this system with water circulated through the radiators on sunny days (Erell and Etzion, 1996).

As with other roof pond systems, waterproofing of the test cell roof had to be done with great care. Subsequent work has shown that this is best resolved by the installation of a PVC lining sloped to a drain to remove rainwater or leaks from the roof pond.

Water Radiator and Cooling Panel

A system combining a water radiator linked to a cooling panel was tested at the Centre for Renewable Energy Sources (CRES), at Pikermi outside Athens (latitude 37°58'N, longitude 23°42'E, altitude 130 m) (Dimoudi *et al.*, 1998). Such a combination may be of interest as a retrofit on buildings of more than one floor. For the test, however, the water radiator and cooling panel were installed on a single-storey test cell with a removable roof (Fig. 4.11). The water radiator was made of heavy-duty steel pipes 3/4" (19 mm) in diameter, with 10 pipes per metre roof width fixed on a steel plate and laid exposed to ambient conditions. The pipes and steel plate were painted white to reflect solar radiation. The cooling panel was in the form of a 120 mm concrete slab with 3/4" (19 mm) steel pipes embedded at 200 mm centres, 5 pipes per metre. A 60 mm layer of thermal insulation was applied between the water radiator and the cooling panel. Two manifolds across the width of the roof connected the pipes of the radiator with pipes in the cooling panel. Water was circulated in a closed loop. The cooling panel absorbed heat from the indoor space, which was then transferred to the radiator to be dissipated to the night sky. During daytime the roof was covered with an insulated panel for protection from solar radiation. The water radiator system installed on the CRES test cell can be seen in Fig. 4.12.

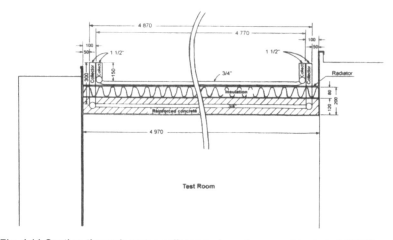

Fig. 4.11 **Section through water radiator and cooling panel tested at CRES.**

Fig. 4.12 View of the test cell roof with the assembled roof component and pipe connections.

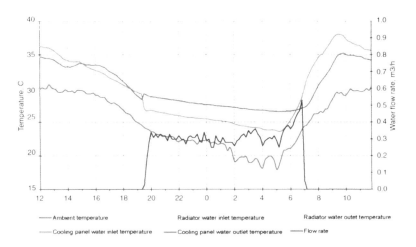

Fig. 4.13 Outdoor air temperature, water temperatures and water flow rate on a day in June for room temperature of 28°C.

The tests included low (0.3 m³/h), as well as medium (0.75 m³/h) and high (1.3 m³/h) flow rates, control of flow regime and control of indoor temperature. It was found that the flow rate required for optimum performance depended on the charge and discharge characteristics of the cooling panel and on the cooling rate of the radiator. This in turn depended on indoor and ambient conditions and on the design of the individual components. Figures 4.13 and 4.14 show results from a period during which the internal temperature of the test cell was kept constant at 28C. The water was circulated between 1900 and 0700 hours. It was observed that for half of the testing period, the water temperature at the outlet of the radiator had risen above that at the inlet by around 0630 hours. This suggested that the optimum period for the operation of the water pump ended at 0630 hours. The opposite effect was observed with the cooling panel. The water temperature at the outlet of the pipes embedded inside the concrete remained below the inlet temperature after 0630 hours on most days. After about 2100 hours the heat flux through the roof became higher than the heat extracted from inside the pipes.

On this system the radiator surfaces exchange heat with the external environment by radiative transfer to the sky (Q_{net} representing the net outgoing longwave radiative flux) and by convective heat transfer (Q_{conv}) to adjacent ambient air layers. The cooling power of the radiator ($Q_{cool,rad}$) is determined by the temperature difference of the water at the radiator's inlet and outlet together with the

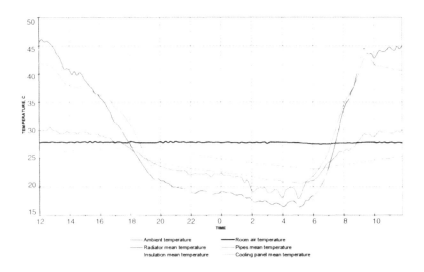

Fig. 4.14 Outdoor air temperature, average test room air temperature and mean values of radiator surface, radiator pipes, insulation layer, and concrete slab temperatures on a day in June for room temperature of 28°C.

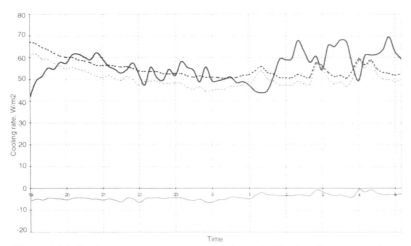

Figure 4.15 Cooling rates on a summer night period for low flow rate and constant T$_{room}$= 28°C

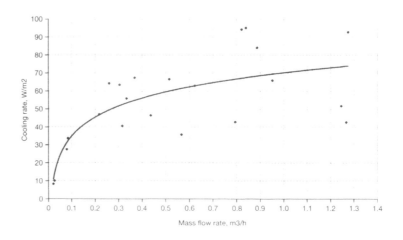

Figure 4.16 Averaged nightly cooling rates against water flow rate.

flow rate. The total cooling rate (Q$_{tot}$) was calculated as the cumulative effect of the net radiative and convective heat exchanges. This varied considerably from day to day. Depending on the temperature of the radiator surface, the radiative and convective mechanisms could be complementary or oppose each other. Calculated rates of these parameters are plotted in Fig. 4.15 for a low water flow regime and room temperature kept constant at 28°C. Over the test period, the radiator cooling rates ranged from 9.5 to 95.8 W/m² with a mean value of 55.9 W/m². The water flow rates ranged from 0.03 to 1.27 m³/h. Figure 4.16 plots the averaged cooling rates against the flow rate. It can be seen that there is an increase of the cooling rate with the flow rate (correlation coefficient R^2=0.45). The water flow rate was found to have significant effect on water temperature differences between inlet and outlet in both radiator and cooling panel pipes, as can be seen in Fig. 4.17 (correlation coefficient R^2=0.94). At low flow rates the temperature difference between the inlet and outlet at the radiator pipes rose to some 6.5K, whereas with high flow rates it decreased to 0.5K.

Overall the experiment showed that this roof system performed effectively. Regulation of the water flow rate is a key issue for effective radiator performance. The system can be retrofitted by laying the water pipes on an existing concrete slab, adding a further layer of concrete, with the thermal insulation and water radiator laid on top. The cost of such construction including labour was estimated at between 105 and 116 Euro/m². A breakdown of costs is shown in Table 4.1.

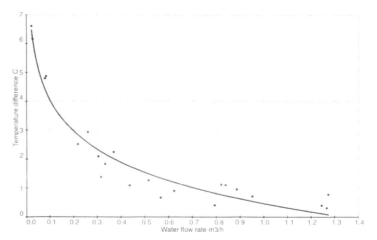

Figure 4.17 Averaged radiator inlet/outlet temperature difference against water flow rate.

TABLE 4.1 COST OF CRES WATER RADIATOR AND COOLING PANEL

DESCRIPTION	COST (EURO)	
1 Steel pipes	718	(910)*
2 Concrete	146.7	
3 Reinforcement for the concrete slab	123	
4 Insulation (8 cm)	132	
5 Radiator exterior metal sheet	132	
6 Labour (3 days)	293.5	
Total	**1545.2**	**(1737.2)**

* Higher price is for galvanised steel pipes.

Water-based Radiators

Two novel radiators were tested on the roof of the CDAUP (Centre for Desert Architecture and Urban Planning) demonstration house, Sde-Boqer (latitude 30°48'N, longitude 35°58'E, altitude 475 m; Fig. 4.18).

"Heliocoil" Radiator

This system made use of a commercially available solar collector panel that is commonly used for swimming pool heating. The panel consisted of thin polypropylene tubes 6 mm in diameter, with an internal diameter of some 4.5 mm. The tubes were connected to a 40 mm diameter manifold, and spaced at 10 mm centres. The radiator panel was 2.3 m long by 1.2 m wide. Polypropylene has an infrared emissivity of some 0.95, which made the Heliocoil collector well-suited for use as a cooling radiator.

Polycarbonate Radiator

This radiator was constructed of a double-skinned polycarbonate panel, attached to 25 mm polycarbonate tubes that served as manifolds at either end of the sheet. The polycarbonate sheet, which is claimed by the manufacturer to be practically opaque to longwave radiation, comprised outer skins approximately 0.5 mm thick, joined by thin internal ribs perpendicular to the outer skin at 10 mm centres. Water flowed between the two skins.

Performance

The cooling output of the two radiators was found to be closely related to the temperature difference between ambient air and radiator inlet. This also correlated with the sky temperature depression (Fig. 4.19). Figure 4.20 shows ceiling and room temperatures on a typical summer day compared with roof pond and ambient air temperatures. The performance of the system was superior to any of those tested on a full-scale building. Using the polycarbonate sheet radiator, mean nightly cooling power averaged over 100 W/m², under typical Sde-Boqer summer conditions (Etzion and Erell, 1998). This cooling output is indicative of the potential of radiative cooling under very clear sky, dry conditions. The cooling output might be lower under less favourable environmental conditions.

The two systems tested provide improvements in terms of efficiency, cost and durability compared with the modified solar flat-plate collectors tested previously.

Efficiency

Both radiators operated with high efficiencies; differences in cooling output were due to the arrangement of the pipes. The polycarbonate sheet was in very good contact with the water, whereas with the

Fig. 4.18 The "Heliocoil" (top) and polycarbonate (left) radiators tested at CDAUP, Sde-Boqer.

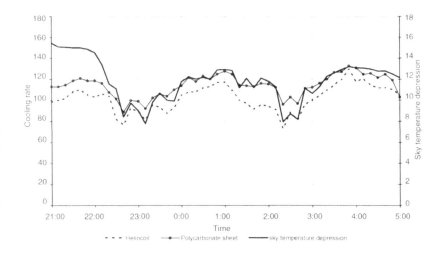

Fig. 4.19 Instantaneous night-time cooling rate with clear sky and sky temperature depression calculated from radiation measurements.

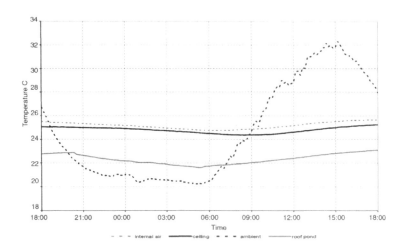

Fig. 4.20   Selected experimental parameters for a typical summer day.

Heliocoil radiator, the pipes comprise approximately 80% of the gross area of the radiator.

Cost

Both radiators cost much less than conventional flat-plate collectors. The Heliocoil radiator cost the equivalent of 120 Euro for a 2 m² unit. The polycarbonate sheet cost approximately 30 Euro per m², so a 2 m² unit could be expected to cost less than 100 Euro if produced commercially.

Buildability and Maintenance

The polypropylene pipes do not corrode or deteriorate by exposure to sunlight, and are not clogged by minerals deposited from the water. The useful service life of such a unit, barring mechanical damage or exposure to sub-zero temperatures when full of water, is indefinite. The polycarbonate (PC) sheet radiator is based on an ordinary PC sheet with a UV protective coating, attached to proprietary manifolds made of the same material. Both radiators are available at almost any length, are very easy to transport and require little expertise to install.

DESIGN CONSIDERATIONS

The performance of a radiative cooling system depends on two types of factor – the properties of the radiator and the operating conditions under which it is run. The graphs included in this section are based on empirical models developed from the experimental work discussed above and from mathematical models for water-based and air-based radiators (Rodriguez and Molina, 1998). The latter are presented in Appendix A and incorporated in the software described in Appendix B. Performance data for radiators are given in Chapter 6.

The Radiative Properties of the Cooling Surface

An ideal radiator should have a high emissivity in the wavelengths between 8 and 13 μm, and be highly reflective in all other parts of the spectrum. Figure 4.21 shows the influence of radiator emissivity on the energy emitted from a hypothetical radiator. Of particular interest is the curve labelled as "0.1+0.9+0.1", which represents the performance of a selective radiator with high emissivity (0.9) in the atmospheric window, and low emissivity (0.1) in wavelengths greater than 13 μm or smaller than 8 μm. The behaviour of this selective surface is more or less similar to a perfect black body surface.

Most materials with the desired properties are applied as thin films deposited on highly reflective aluminum foil. Berdahl (1984), for example, experimented with MgO and LiF backed with aluminium foil, and reported favourable spectral properties. Solid polished MgO surfaces also have a very high solar reflectivity, opening the possibility of using the material as a radiator capable of producing net cooling even during daytime. The decision whether to use a selective coating on the radiator should be made considering the following issues:

1. When the water inlet temperature is above that of ambient air, its effect on cooling output is greater than the selectivity of the radiator (Fig. 4.22). When the inlet water temperature is well above ambient, the performance of a grey surface is better than that of a selective one, because radiation in wavelengths outside the atmospheric window is stronger. Therefore, if the temperature of the radiator is expected to be above ambient, a dark grey (near black) finish is preferable to a selective coating. Spectral differentiation may be advantageous only when the radiator temperature is expected to be below ambient during operation.

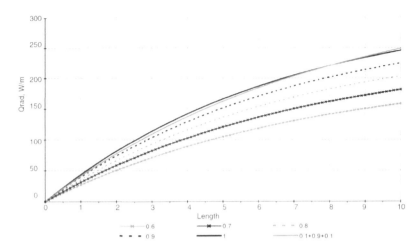

Fig. 4.21 Influence of radiator emissivity.

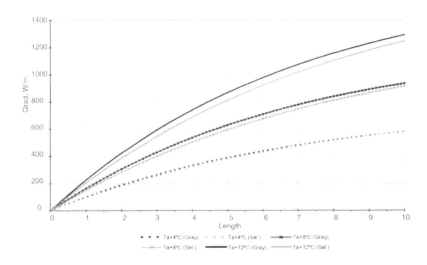

Fig. 4.22 Influence of the selectivity (sel.) of the radiator.

2. The application of selective emitting materials to radiative cooling on a commercial scale requires a means of maintaining the optical properties of the surface after prolonged exposure to the environment. Selective coatings need to be kept clean to preserve their advantage over conventional radiators. In addition, they must be kept dry – the formation of dew on the radiator eliminates any spectral selectivity the surface may have. These practical considerations may therefore rule out the application of selective coatings to roof-mounted radiators, at least until the problems raised here are resolved.

The solar absorptivity of the plate is only important where the radiator is also used as a solar collector, for example, to provide domestic hot water or swimming pool heating. It is of minor importance when the system is designed for nocturnal operation only. Where the heat exchange fluid is not circulated during daytime, it may be preferable to paint the radiator in a light colour, thus providing it with a low solar absorptance to reduce solar gains during the daytime. If the radiator is not in direct thermal contact with the building, the colour is irrelevant, since even a black radiator cools down very rapidly near sunset, and cooling operation is not delayed (Erell and Etzion, 1992).

Windscreens

In favourable environmental conditions, longwave radiation to the night sky can lower a radiator's surface temperature below that of the ambient air. When this happens, however, the ambient air acts as a source of heat gain by convection, thus counteracting the radiative heat loss. The point at which radiative cooling and convective heat gain exactly cancel each other out is known as a radiator's *stagnation* temperature.

When the fluid entering a radiator is at a temperature well above that of the ambient air, most of the radiator surface is also likely to be above that temperature and convection will contribute to heat loss from the radiator (Fig. 4.23). In such cases, wind improves radiator performance, and the addition of a windscreen is counterproductive.

Windscreens are only required if the radiator must be cooled below ambient air temperature. In this case, the stagnation temperature may be lowered by creating a layer of still air above the radiator that is prevented from mixing with the surroundings. This may be done by placing an open honeycomb covering, preferably made of a

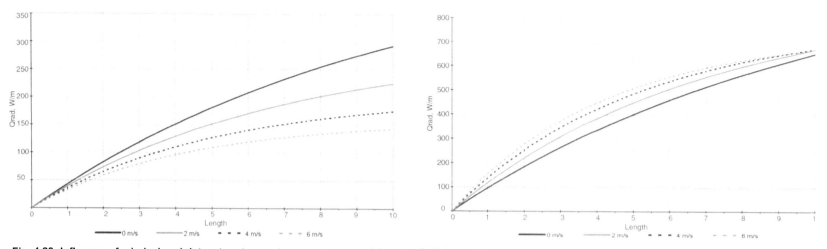

Fig. 4.23 Influence of wind when inlet water at same temperature as ambient air, T_a (left graph), or higher than ambient air (right).

material that is highly reflective in the infra-red spectrum (Martin, 1989). Such a covering would allow radiation to be emitted from the radiator, since it would be open at the top, while restricting the motion of air near the radiator surface. A simpler approach, investigated by several researchers (Addeo et al., 1980; Berdahl et al., 1983), is to cover the radiator with glazing that is transparent in the infra-red spectrum, particularly between 8 and 13 μm. Such glazing would transmit infra-red radiation emitted by the radiator, but prevent air sealed beneath it from mixing with the surrounding air. Glass and other materials with adequate mechanical strength and chemical stability are opaque over this spectral range. Among polyolefin plastics, polyethylene film manufactured without UV inhibitors may be suitable for this purpose. A very thin (2-μm or 0.05-mm thick) polyethylene film has infra-red transmissivity of about 75%. While this material is quite cheap, it rips quite easily and deteriorates rapidly when exposed to solar radiation. Also, lacking rigidity, it must be stretched tightly over a supporting grid or reinforced by a web of nylon or another similar fibre to provide tensile strength. If it is not held in tension, it flutters in the wind, producing air motion in the space between the windscreen and the radiator, thus losing most of its insulating effect. A number of research projects have been undertaken, directed towards the development of an improved infra-red transparent glazing based on polyethylene.

These projects have focused on the possibility of coating a polyethylene sheet with a reflective pigment designed to reduce solar gains and increase the longevity of the material. The use of ZnO as a reflective pigment, for example, produced a polymer film that showed no signs of physical or spectral degradation after 1200 hours of exposure to sunlight under accelerated testing procedures (Brookes, 1982).

The temperature on the surface of the windscreen will often drop to the dew point of the surrounding air. Once moisture begins to condense on the surface, the windscreen becomes, in effect, the primary radiator, albeit one exposed to the very effects of convection it was designed to prevent. This is a common occurrence, even in very dry locations, since the clear atmosphere of such regions promotes nocturnal radiative cooling. The accumulation of dust or airborne pollutants can reduce the infra-red transmissivity of the glazing significantly. This problem is exacerbated by the presence of dew on the exposed surface, since dust particles deposited on a wet surface are retained, rather than being blown away by the wind. Frequent cleaning is required.

Tilt Angle of Radiator

A radiating surface positioned horizontally has the highest net radiative cooling power. Tilted surfaces are exposed to the warmer

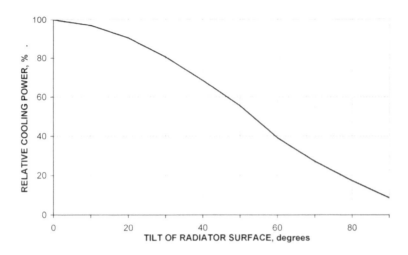

Fig. 4.24 Effect of tilt angle on net radiative cooling of exposed surface. (Derived from Clark and Berdahl, 1980)

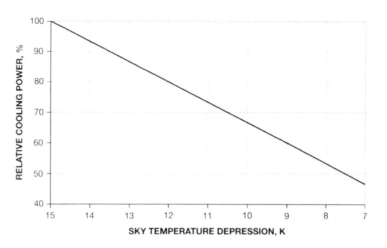

Fig. 4.25 The effect of a reduction in the sky temperature depression on the net radiative heat loss of surface with _e_=0.9 at ambient air temperature.

regions of the sky near the horizon, and also to the ground, which is generally warmer than the sky. Practical considerations, such as roof drainage or architectural design may, however, require the radiator surface to be tilted. Simulation studies have shown that small tilt angles of up to 10°, sufficient for water drainage, have little effect on the net radiative cooling output; however, a slope of 30°, required for tiled roofs, would result in a reduction in net cooling of 20% (Fig. 4.24; Clark and Berdahl, 1980).

Radiator Insulation

In solar collectors, the purpose of applying thermal insulation behind the collecting surface is to reduce unwanted heat losses, primarily by convection but also by radiation. In the case of nocturnal radiators, however, the objective is to maximise losses. Back insulation is therefore desirable only if the operating temperature of the radiator is expected to be well below that of the ambient air.

If the radiator is also being considered for secondary use as a solar collector during winter, conventional thermal insulation should be applied. (Normally, insulation is between 10- and 20- mm thick, assuming conventional materials such as polystyrene or polyurethane foam, with a conductivity of 0.03–0.04 W/mK).

Operating Parameters

The cooling system as a whole should be designed to control the following factors:

- The *radiator inlet temperature* should ideally be as warm as the warmest part of the building to be cooled. The means of transferring energy accumulated in the thermal mass of the building to the radiator is a critical part of a radiative cooling system.
- The *mass flow rate* should be controlled to achieve a fairly flat temperature profile along the length of the radiator. In dry locations where the sky temperature depression is generally about 10–20K, a reduction of even 1°C in the average temperature of the radiator results in a loss of 10–15% of the net radiative cooling output (Fig. 4.25).

The following sections introduce the design paramenters for water-based and air-based radiators more spcifically. Table 4.2 at the end of this chapter provides summary design guidelines for both types of cooling radiator.

WATER-BASED RADIATORS

Radiator Properties

An unglazed flat-plate solar collector can be used as a cooling radiator. A typical flat-plate collector, shown in Fig. 4.26, consists of parallel riser pipes connected to main distributors (headers) at the inlet and outlet. In order to achieve high fluid temperatures, each of the pipes is in turn attached to a fin, designed to concentrate solar energy from a relatively large surface area onto a small volume of water. The fins are sometimes joined together to form a continuous plate. The mass flow rate in a solar collector is generally very slow – about 0.003–0.05 kg/m²/s, corresponding to a water velocity of about 0.5–10 cm/s.

The design of a solar collector is concerned with parameters such as the diameter and spacing of the risers, the thickness and conductivity of the pipes and fins (or cover plate) and the quality of the thermal bond between them. The overall performance of a collector is characterised by a collector efficiency factor and a fin efficiency factor, which relate the actual useful energy gain of a collector to the useful gain that would result if the collector absorbing surface had been at the local fluid temperature (Duffie and Beckman, 1991). In the case of a flat-plate collector modified for use as a cooling radiator the following considerations apply:

1. High fin efficiency is desirable only if the radiator operates at temperatures well above that of ambient air; if the fluid temperature is lower than that of ambient air, high fin efficiency increases convective heat gains, thus reducing the net amount of energy given off by the radiator.
2. Fin efficiency may be improved (assuming the radiator is warmer than the air) by using thicker, more conductive fins, and by reducing the distance from the edge of the fin to the tube with the fluid; Unlike a solar collector, fins might ultimately be dispensed with altogether, leaving a continuous emitting surface. If fins are retained, the minimum thickness required for structural reasons is also sufficient to satisfy thermal requirements (for metal plates, this is about 1–2 mm). The effect of fin width (W) is shown in Fig. 4.27. For any value of the radiator's heat loss coefficient (U_L), itself affected by convective exchange, a narrower fin is more efficient.

The effect of collector efficiency on the cooling output of a hypothetical radiator is illustrated in Fig. 4.28. For the given radiator design and environmental conditions, an increase in collector effi-

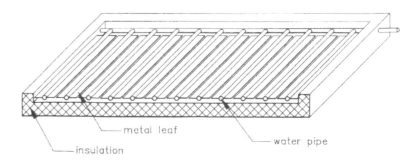

Fig. 4.26   Axonometric view of typical pipe-and-fin flat plate solar collector that can be used as a cooling radiator.

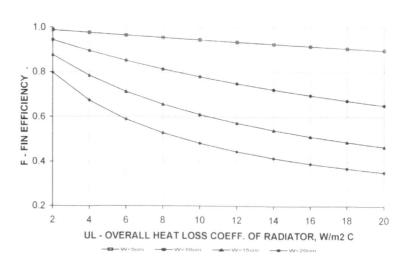

Fig. 4.27   Fin efficiency as a function of heat loss coefficient for different fin widths.

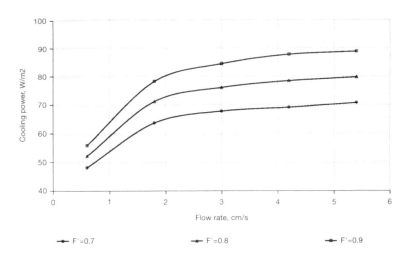

Fig. 4.28   Cooling output as a function of flow rate for different overall effect of the overall efficiency factors (*F'*).

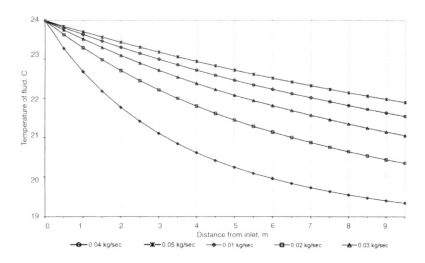

Fig. 4.29   Longitudinal temperature profile of a cooling radiator as a function of mass flow rate.

ciency from 0.7 to 0.9 may result in an increase in cooling output of approximately 15% at low flow speeds (less than 1 cm/s), and up to 25% at high flow velocities (3 cm/s or more). The difference in absolute terms is even greater, since the cooling rate is slightly greater at high flow rates. Optimisation of the collector design should therefore focus on operation at relatively high flow velocities of the fluid (3 cm/s or more).

Operating Parameters

The effects of operating parameters such as the inlet temperature and the mass flow rate are shown in Figs 4.29 and 4.30 for a hypothetical radiator 1 m wide, with overall efficiency (*F'*) of 0.8. The environmental conditions assumed were an ambient air temperature of 22°C, sky temperature depression of 15K, and little wind. Under these conditions, an unglazed, exposed radiator would have a convective heat loss coefficient (U_c) of about 10 W/m²K. Figure 4.29 shows the effect of altering the flow rate on the longitudinal temperature profile. For a specific radiator length, any desired outlet temperature may be attained, within the constraints of the environmental conditions, by manipulating the flow rate.

Figure 4.30 shows the effect of altering the flow rate on the cooling output at several inlet temperatures. Increasing the mass flow rate reduces the temperature difference between radiator inlet and outlet, resulting in an increase in the mean surface temperature of the radiator. This, in turn, results in increased radiative cooling, under

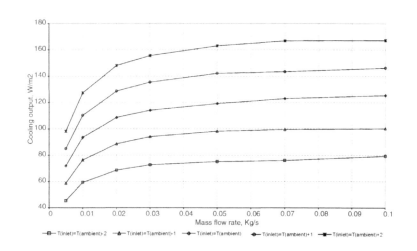

Fig. 4.30   Radiator cooling power as a function of mass flow rate for different fluid temperatures at radiator inlet.

all environmental conditions. When the radiator is warmer than the ambient air, increasing the flow rate also increases the cooling output through convection; if the radiator is cooler than the ambient air, convective heat exchange tends to counteract the effects of radiation. Increasing the fluid flow does, however, have diminishing returns. As the flow rate increases, the surface temperature of the radiator approaches that of the fluid at the inlet, which in a cooling radiator is the theoretical limit. The practical limit takes account of the power required to operate the pump as well as the heat transfer processes occurring at the radiator surface.

Finally, it is worth noting the following:

• *heat exchange between the tube walls and the fluid* is improved by a turbulent flow regime. However, in most collector designs the flow speeds noted above result in laminar flow. An increase in flow rate to force a turbulent regime would not increase significantly the thermal performance and would substantially increase the energy needed to pump the water;

• *a round pipe section is not essential*; other considerations, such as the sky view factor of each pipe, or simplicity of construction, may recommend a radiator with rectangular pipes;

• *optimisation of the hydraulic design* is required so as to attain the flow rate calculated to maximise cooling output, with a minimum investment of electricity to operate the pump;

• *the length of the radiator* may be adapted to the geometry of the roof on which it is installed; the flow velocity and mass flow rate are limited by hydraulic constraints such as the diameter of the manifold pipe or the operating parameters of the pump; thus, there is no advantage to be gained, for instance, from dividing a long radiator into several shorter ones of equal cross-section.

AIR-BASED RADIATORS

Air radiators can be more variable than water radiators in their geometric configuration. The simplest air radiator is a rectangular air channel between a radiating plate and a back surface. Because of the low heat transfer coefficient between the plate and the working fluid, air radiators have a lower efficiency than water-based radiators. Some of the design parameters follow the same trends as those described for water radiators, but there are also some differences. These are discussed below.

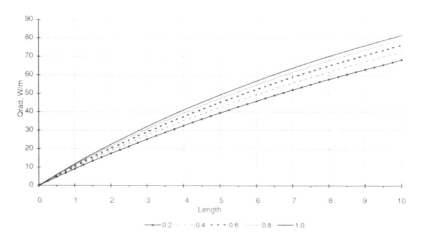

Fig. 4.31 Influence of the interior emissivities in the air channel.

Air Channel Thickness

For a given air flow rate, narrower channels perform better because they tend to increase the heat transfer between plate and air (Fig. 4.31). Typical values for the air channel thickness may range between 10 and 20 mm. Narrow channels also increase friction, and are thus restricted in length.

Air Mass Flow Rate per Metre of Width

An optimum condition is to set a mass flow rate that results in a fully turbulent regime. For instance, for a 20 mm rectangular channel the recommended values for the air flow rate are in the range 360–400 kg/hM (Fig. 4.32). For other channel geometries, an air velocity around 7 m/s should be aimed at.

Plate Thickness

Contact between the plate and the circulating air inside is over the whole plate surface. Thus there is no fin effect influencing performance. As with water-based radiators the minimum thickness of the radiator surface that satisfies structural requirements is also satisfactory for thermal performance.

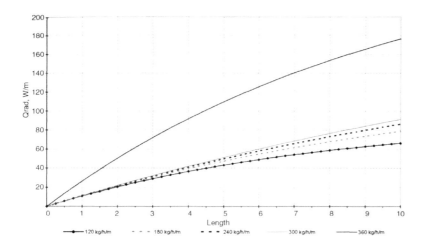

Fig. 4.32 Influence of the air mass flow rate.

Operational Considerations for Air Radiators

Since air as the working fluid has little thermal capacity the cooling produced using air radiators needs to be used immediately of or stored in a specially designed storage system. Architectural integration is more difficult than for other applications.

The control strategy for air radiators is similar to that of water radiators, but in this case the threshold energy for starting the system is higher, and thus the total cooling potential lower.

References

Addeo, A., Nicolais, L., Romeo, G., Bartoli, B., Coluzzi, B. and Silvestrini, V. (1980) Light selective structures for large scale national air conditioning. *Solar Energy,* 24: 93–98.

Berdahl, P. (1984) Radiative cooling with MgO and LiF layers. *Applied Optics,* 23(3): 370–372.

Berdahl, P., Martin, M. and Sakkal, F. (1983) Thermal performance of radiative cooling panels. *International Journal of Heat Mass Transfer,* 26(6): 871–880.

Brookes, J.R. (1982) *Development of Radiative Cooling Materials.* Final Technology Report: FY 1980–1981, DOE Contract No. DE-FC03-80SF11504.

Clark, E. and Berdahl, P. (1980). Radiative cooling: resource and applications. In Miller, H. (Ed.), *Passive Cooling Handbook.* Berkley CA, pp. 177–212.

Dimoudi, A., Sutherland, G., Androutsopoulos, A. and Vallindras, M. (1998) *Development and Testing of a Prototype Radiative Cooling Component Using a Water Radiator Linked to a Cooling Panel.* Roof Solutions for Natural Cooling. Final Report. European Commission Joule Programme.

Duffie, J.A. and Beckman, W.A. (1991) *Solar Engineering of Thermal Processes.* John Wiley & Sons, New York.

Erell, E. and Etzion, Y. (1992) A radiative cooling system using water as a heat exchange medium. *Architectural Science Review,* 35(2): 39–49.

Erell, E. and Etzion, Y. (1996) Heating experiments with a radiative cooling system. *Building and Environment,* 31(6): 509–517.

Etzion, Y. and Erell, E. (1989) A hybrid radiative–convective cooling system for hot-arid zones. In *Clean and Safe Energy Forever.* Proceedings, ISES Solar World Congress, Kobe, Japan.

Etzion, Y. and Erell, E. (1991) Thermal storage mass in radiative cooling systems. *Building and Environment,* 26(4): 389–394.

Etzion, Y. and Erell, E. (1998) Low-cost long-wave radiators for passive cooling of buildings. *Architectural Science Review.*

Givoni, B. (1977) Solar heating and night radiation cooling by a Roof Radiation Trap. *Energy and Buildings,* 1: 141–145.

Givoni, B. (1994) *Passive and Low Energy Cooling of Buildings.* Van Nostrand Reinhold, New York.

Martin, M. (1989) Radiative cooling. In Cook, J. (Ed.) *Passive Cooling.* The MIT Press, Cambridge MA.

Rodriguez, E.A. and Molina, J.L. (1998) *Modelling of water-based and air cooling radiators. ROOFSOL Project Task 2 Final Report .* European Commission Joule Programme.

TABLE 4.2 SUMMARY DESIGN CONSIDERATIONS FOR COOLING RADIATORS

MOVABLE INSULATION	WATER COOLER THAN AMBIENT	WATER WARMER THAN AMBIENT	AIR
The radiator is insulated from the environment when heat exchange is not required.	*The radiator is always exposed. Water is circulated as a heat exchange fluid and is cooled to below ambient temperature.*	*The radiator is always exposed. Water is circulated as a heat exchange fluid, but the radiator is usually warmer than ambient temperature.*	*The radiator is always exposed. Air is circulated as a heat exchange fluid.*

Parameters to consider

MOVABLE INSULATION

- Roof geometry
- Mass of roof

WATER COOLER THAN AMBIENT — Parameters to consider

- Radiator material & colour
- Pipe diameter, length & spacing
- Back insulation
- Windscreen
- Storage mass
- Water flow rate

WATER WARMER THAN AMBIENT — Parameters to consider

- Radiator material & colour
- Pipe diameter, length & spacing
- Back insulation
- Windscreen
- Storage mass
- Water flow rate

AIR — Parameters to consider

- Radiator material & colour
- Thickness & length of air space
- Back insulation
- Windscreen
- Storage mass
- Air flow rate

Recommendations

MOVABLE INSULATION

- Radiating roof must be rectangular. Dimensions are limited by size of insulation boards and mechanical considerations.
- Insulation should be stacked over a service space such as a garage.
- Thermal mass may be provided by water bags, a roof pond or structural concrete.

WATER COOLER THAN AMBIENT — Recommendations

- Dark if used as backup heating system, otherwise light.
- Pipe diameter as appropriate for required flow rate. Length suited to roof dimensions. Pipes as closely spaced as possible.
- Back insulation and windscreen required in order to achieve temperature lower than ambient.
- Storage mass either water reservoir or concrete cooling panels.
- Water flow rate relatively slow to achieve low outlet temperatures.

WATER WARMER THAN AMBIENT — Recommendations

- Dark if used as backup heating system, otherwise light.
- Pipe diameter as appropriate for required flow rate. Length suited to roof dimensions. Pipes as closely spaced as possible.
- Back insulation and windscreen hinder cooling.
- Storage mass either water reservoir or concrete cooling panels.
- Water flow rate relatively high to maintain radiator warm and provide maximum cooling.

AIR — Recommendations

- Highly conductive metal sheet.
- Air gap 1–2 cm. Length limited by friction and power of fans.
- Back insulation and windscreen required in order to achieve temperature lower than ambient.
- Storage mass essential: rocks or concrete cooling panels.
- Air speed less than 7 m/s: much lower air speeds required to supply air substantially cooler than ambient.

5 PLANTED *('GREEN')* ROOFS

INTRODUCTION

Planted or "green" roofs have a layer of vegetation growing on a specially designed substrate over a flat or sloped roof structure (Fig. 5.1). A planted roof may contribute to the following:

- replacing depleted vegetation in urban areas, thus fostering local microclimatic improvements;
- reducing conductive heat gains through the roof structure;
- more stable indoor temperatures.

Unlike roof ponds and radiators, which have had few applications in Europe, green roofs are gaining in popularity (Figs 5.2 and 5.3). For example, it is claimed that in Germany some 7% of new flat roofs are planted (Kohler *et al.*, 2002), and there are applications in most other European countries.

SYSTEMS AND COMPONENTS

The main layers of a planted roof construction are:

1. planting
2. substrate
3. waterproofing layer
4. insulation layer
5. roof structure.

A distinction is often made between *intensive* planted roofs (roof gardens), which require considerable maintenance (Fig. 5.4), and *extensive* planted roofs (*ecological green roofs*) that comprise a thinner layer of soil and are designed to be self-sustaining (Fig. 5.5).

Planting

Plants are commonly chosen to suit the local climate and the structure of the roof. Local species are likely to require less maintenance. On roofs with a slope of 8° and above, the use of turf is recommended.

Substrate

The thickness of the soil substrate used may vary from as little as 10 mm to over 2000 mm. Shallow configurations (below 200 mm)

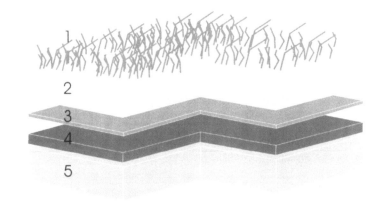

Fig. 5.1 Planted roof components.
1. Planting, 2. substrate, 3. waterproofing layer, 4. thermal insulation, 5. roof structure

Fig. 5.2 Westminster Lodge at the Architectural Association School's Hooke Park site, Dorset, England.
(Architects: E. Cullinan Architects)

Fig. 5.3 Earth-berming and roof planting on northern side of dwellings at Hockerton Housing Project, Nottinghamshire, UK.
(Architects: Brenda and Robert Vale)

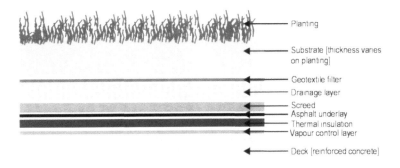

Fig. 5.4   Typical composition an intensive planted roof.

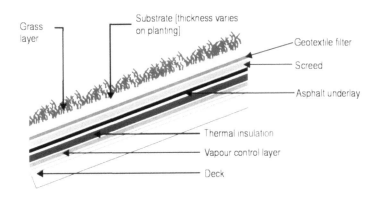

Fig. 5.5   Cross-section of an extensive planted roof.

are typical of extensive planted roofs. Roof gardens require thicker substrates, on which trees, grass and other plants can grow. Heavy soils can be lightened by adding lightweight clay granules.

Waterproofing Layer
This is usually in the form of three coats of mastic asphalt to a total thickness of 30 mm on horizontal surfaces and 20 mm on vertical.

Thermal Insulation
The thermal insulation materials specified for planted roofs should be water resistant.

Roof Structure
The roof structure is most commonly of reinforced concrete. It must be able to support all the layers of a planted roof. The structural calculations should account for dead loads of vegetation, insulation, waterproofing and substrate in a saturated state, and live loads on roof gardens from people, furniture and equipment. Typical loading values associated with planted roofs are given in Table 5.1.

TABLE 5.1	PLANTED ROOF STRUCTURAL LOADS
Top soil	16–20 kg/m^2
Sand	20–22 kg/m^2
Gravel	16–18 kg/m^2
Standard soil	7–9 kg/m^2
Aerated soil (diameter 8–16 mm)	3 kg/m^2
Turf	5 kg/m^2
Typical weights of planted roofs	
Lightweight planted roof	25–30 kg/m^2
Extensive planted roofs	65–150 kg/m^2

ENVIRONMENTAL DESIGN PRINCIPLES

Plants affect the microclimate near the roof in several ways:

1.   Radiative, sensible and latent heat fluxes are spatially variable within the vegetative canopy, so that:
     - penetration of shortwave solar radiation is reduced;
     - longwave radiation from the surface to the atmosphere is intercepted;
     - wind speed near the surface is reduced, and the effects of advection lessened;
     - surface run-off of water is reduced substantially.
2.   Energy storage follows two separate mechanisms:
     - sensible heat is stored in the roof structure and in the soil, as well as in the plants;
     - biochemical energy is stored in the plants.
3.   Latent heat exchange occurs to a large extent due to evapotranspiration from plant leaves.

A planted roof has a cooling and cleansing effect on the adjacent air layers thus contributing to the improvement of urban microclimates. The shading effect and evapotranspiration of the plants should have a cooling effect on the substrate below; however, the plant foliage may inhibit heat dissipation. The thermal capacity of the substrate can provide heat storage that may further reduce and delay the effect of heat gains from solar radiation and ambient air (Fig. 5.6). The contribution of the planting to the thermal performance of the roof thus mainly depends on:

- the density of the foliage;
- the water content of the substrate;
- the composition, density and thickness of the substrate.

Reductions in air-conditioning loads have been reported in the literature (McPherson *et al.*, 1988, 1989; Wong *et al.*, 2003a,b). However, most of these studies have estimated the energy savings by comparison with uninsulated roofs. In cool climates, roofs, whether planted or not, must be provided with a substantial thermal resistance to control winter heat losses. On a roof that is already well insulated, the soil substrate and planting are unlikely to have a noticeable cooling effect in summer.

ADVANTAGES AND DISADVANTAGES

Planted roofs can provide environmental, visual and technical benefits. These include:

- microclimatic effects on the urban environment resulting in reduction of the heat island effect;
- absorption of carbon dioxide, air and water pollutants and dust;
- redressing ecological diversity in cities;
- increase in life expectancy of roof membranes (owing to protection from solar radiation);
- alleviating heat stress due to temperature fluctuations on roofing;
- thermal and acoustic insulation;
- visual appeal.

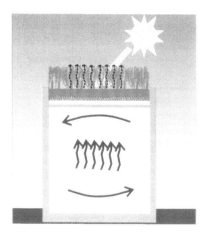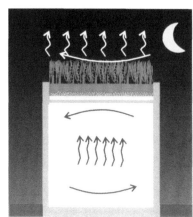

Fig. 5.6 Planted roofs, left: daytime, right: night.

Intensive and extensive planted roofs differ in:

- the choice of plants;
- the thickness and composition of the substrate;
- complexity;
- maintenance requirements.

Extensive planted roofs are designed to be self-sustaining, thus requiring minimal maintenance. The growing medium is typically quite shallow, often no more than 50 mm. There is generally no need for special strengthening of the supporting structure. These types of planted roofs are suitable for slopes of 0–30°. Planting is selected for its capacity to endure extreme temperatures and long periods without water irrigation, and to grow on shallow substrates. Intensive planted roofs, or roof gardens, require a growing substrate of 200 mm or more and a watering system. They require more intensive management and impose a greater loading on the supporting structure.

PRIVATE RESIDENCE AT PANORAMA
Panorama, Thesaloniki, Greece (Figs 5.7–5.9)

Timber-frame sloped roofs planted to harmonize with surrounding landscape.

Key Features

- Water collected from roof at reservoir and used to water plants.
- Watering of roof after sunset to help cool adjacent air expected to contribute to cooling of building's thermal mass in the evening and night.
- Heat dissipation by stack effect through openings in higher part of roof.

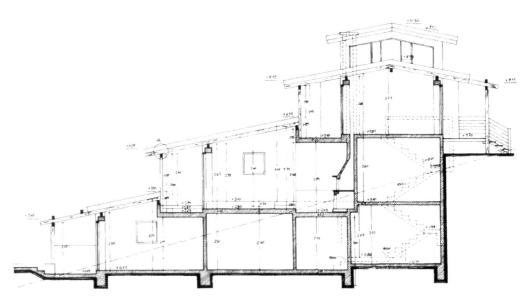

Fig. 5.8 Section. House at Panorama

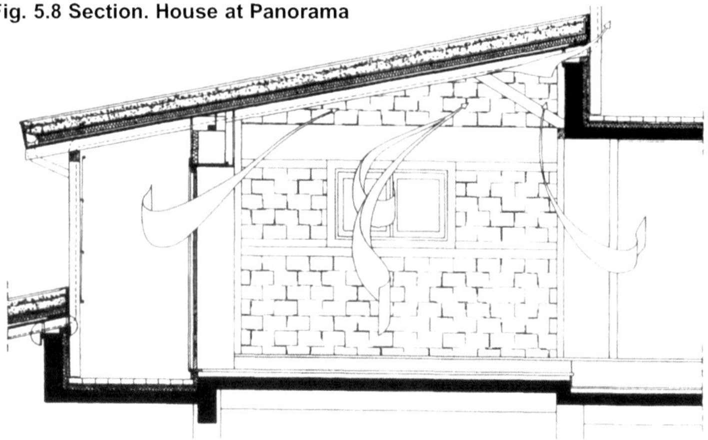

Fig. 5.7   House at Panorama under construction.

Client & Architect: Elli Georgiadou
Building type: Two-storey detached house
Date completed: 1990s
Floor area: 114 m²

Fig. 5.9   Detail.

Roof construction

- Planting of local wild species.
- Soil layer over waterproofing membrane with soil-fabric layer.
- Waterproofing and drainage membranes over insulation.
- 75 mm thermal insulation.
- 30 mm air gap for ventilation.
- Timber roof structure.

APPLICATIONS

ESCUELA TECNICA DE INGENIEROS AGRONOMOS
Madrid, Spain

This application comprised 400 m² of planting organised in a series of parcels that were laid over an existing roof of conventional concrete slab construction at the School of Agricultural Engineers' Fitotecnia building in Madrid. Figure 5.10 shows the installation of the PVC waterproofing membrane over the existing roof. Figure 5.11 shows a view of the planted roof in the spring, six months later. The parcels varied in composition based on different choices of the following parameters:

- *Waterproofing*: half of the terrace has a waterproofing layer based on asphalt tar and the other half has PVC membrane.
- *Drainage materials*: some parcels have a membrane of polystyrene nodules between permeable textile sheets; other parcels have a perforated plate of expanded polystyrene or absorbent rock wool panels.
- *Organic substrate*: forest waste (basically pine) with added organic matter from purified mud (10%) in a 50 mm layer; mineral substrate of 30 mm.
- *Inert substrate*: placed over the organic substrate as protection; lapilly or expanded clay were used.
- *Plants: Sedum album* was the only type used; this is abundant in the area, needs no watering or maintenance, can sustain extreme temperatures and long periods without rainfall, its roots can grow in substrates of 5–8 cm, and it maintains its aesthetic effect throughout the year.

Figure 5.12 shows a section of the layers on some of the planted roof parcels. Figures 5.13 and 5.14 illustrate a configuration that is available commercially in Spain as Azotea Aljibe Ecológica ("ecological cistern roof"). This has a water pond under the planting; the water is replenished from rainfall.

Results of measurements taken in different parcels are shown in Figs 5.15–5.18. Some of the parcels were variants of the Azotea Aljibe system whilst others had been left without plants for comparison.

Fig. 5.10 Waterproofing in preparation for the laying of the planted roof parcels.

Fig. 5.11 Planted roof six months after planting.

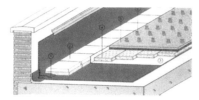

Fig. 5.12. Sections of the planted roof.
1. Waterproofing membrane; 2. corridors of porous concrete fixed to expanded polystyrene; 3. water retainer membrane; 4. drainage layer; 5. substrate; 6. plants.

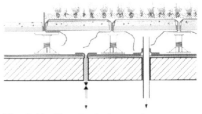

Fig. 5.13 Detail of the ecological cistern roof ("Azotea Aljibe Ecológica").
From the bottom to the top: 1. Existing flat roof; 2. feltemper membrane – made of a textile; 3. PVC waterproofing layer; 4. water pond – with the supports of the upper layers embedded, and the absorbent textile membrane (made of Feltemper 300) falling into the water; 5. Losa Filtron porous concrete and insulation tile; 6. substrate; 7. plants – *Sedum album*.

Fig. 5.14 Construction detail of the "Azotea Aljibe Ecológica" showing the water absorbent layer below the "Losa FILTRON" tile.

Figure 5.15 compares the measured temperatures at the upper layers of parcels P05, P06, P07, P09, P10 and P12, all fitted with variants of the Azotea Aljibe Ecológica, with those of parcels 4, 44 and 50, which are unplanted. It can be seen that on a July day with ambient air temperature approaching 30°C, all three unplanted parcels reach peak temperatures in excess of 40°C whereas the planted parcels had peaks below 30°C. It is notable also that at parcels P7 and P9 the planting displayed peak temperatures below ambient air temperature.

Also of interest is the comparison in Fig. 5.16 between parcel P17, which has the specification shown in Fig. 5.12, and parcel P20 which had no planting. At a depth of 1.5 cm below the surface, the July temperature of parcel P20 reached a peak at 1200 hours, 5K higher than parcel 17. However, the minimum temperatures at P20 were some 4K lower than those of P17 at 0500–0600 hours. At a depth of 5.5 cm the maximum temperature of parcel P20 was 2K higher and its minimum 2K lower than the corresponding values of parcel P17. At a depth of 16 cm, the temperatures in the two parcels coincided, owing to the thermal capacity of the substrate.

The heat flux measurements at 16 cm depth showed heat gains between 1300 and 2400 hours with a maximum of 5 W/m² at 1700–1800 hours. The heat losses over the rest of the 24-hour period have a similar profile, suggesting an overall energy balance close to zero.

Figures 5.17 and 5.18 show hourly temperatures and heat fluxes at different depths for parcel P17 on one summer day and one winter day.

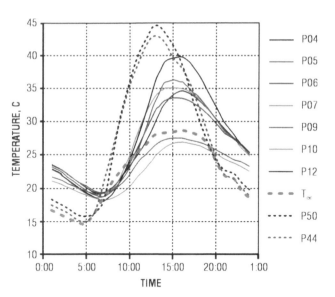

Fig. 5.15   Comparisons on a typical July day of upper layer temperatures of parcels with Azotea Aljibe Ecologica (parcels P05, P06, P07, P09, P10, P12) and reference parcels P4, P44 and P50; T_{ext} is the ambient air temperature.

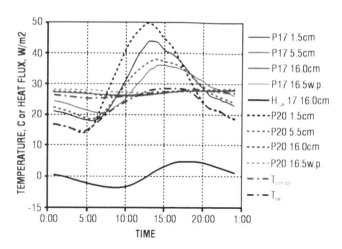

Fig. 5.16   Temperatures and heat fluxes in parcels P17 and P20 in July.

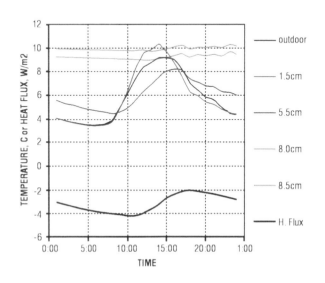

Fig. 5.17   Mean hourly temperatures and heat fluxes at different depths for parcel P17 in January.

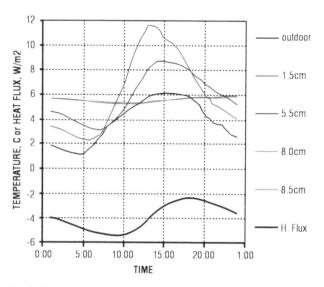

Fig. 5.18 Mean hourly temperatures and heat fluxes at different depths for parcel P17 in July.

DESIGN CONSIDERATIONS AND PERFORMANCE

Introduction

A mathematical model was developed in the context of the ROOF-SOL project to study the thermal processes occurring in planted roofs, and to evaluate their cooling potential in summer (Palomo Del Barrio, 1998). The mathematical basis of the model is summarised in Appendix A and results from its application are presented here. The model combines components representing the soil, the structural roof and the plant canopy (Fig. 5.20).

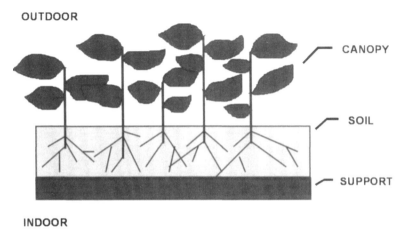

OUTDOOR

CANOPY

SOIL

SUPPORT

INDOOR

Fig. 5.20   Schematic of planted roof model components.

These are characterised by the following assumptions:

- the *soil* is assumed to be a porous medium incorporating minerals, liquid water and water vapour; heat transfer is by conduction, convection and latent heat transfer by vapour diffusion;
- the *structural roof* is a homogeneous layer of solid material with constant physical properties and heat transfer by conduction;
- the *plant canopy* absorbs solar radiation in leaves and soil, is subjected to radiative exchanges between leaves, sky and soil, convective heat transfer between leaves, canopy air and ground; evapotranspiration in leaves, evaporation or condensation of water at the soil surface; and convective heat and vapour transfer between the air in the canopy and ambient air.

A set of studies were performed with this model for a hypothetical planted roof, using weather data for Athens, Greece. The daily ambient air temperature in the period encompassed by the data was in the range 24–34°C, relative humidity varied in the range 40–75% and global solar radiation on a horizontal surface was of some 900 W/m² at midday. Other inputs to the model were the long-wave radiation flux from the sky, wind speed and wind direction. The air temperature in the building was assumed constant at 25°C.

The parameters identified as having a bearing on the thermal performance of the planted roof were: the density, thickness and moisture content of the soil; the density and geometric characteristics of the foliage; and air motion through the canopy. The effects of each of these parameters are summarised below.

Soil Substrate

The selection of soil for a planted roof will affect not only the development of plants, but also the thermal and hydrological performance of the roof. The *apparent density* of a soil is determined not only by the density of its constituent particles, but mostly by the grain size distribution. A well-sorted soil, consisting of particles that are all of similar size, will tend to have a relatively larger volume of air spaces, and will thus be lighter than a poorly sorted soil, with particles having a range of sizes. Concurrently, smaller particles, such as clay, will result in relatively denser soils than larger particles, such as sand. All else being equal, a dense soil will allow a larger heat flux through the roof than a lighter soil (Fig. 5.21). If, as is usually the case, the weight of soil on a roof is limited by structural considerations, then a light soil not only provides better insulation per unit depth but it also allows a deeper layer of soil to be provided for a given dead weight.

The *soil moisture content*, defined as the percentage volume of a moist soil occupied by water, has an extremely important effect on its thermal properties, as well as on evaporation occurring at its surface and on moisture availability to plants. In a dry soil, thermal conductivity increases with soil moisture, because the addition of water increases the thermal contact between grains, and because water replaces air, which has a much lower conductivity. As the soil becomes waterlogged, the increase in thermal conductivity gradually levels off. The addition of water also increases the heat capacity of a soil, which is an indication of its capacity to store energy.

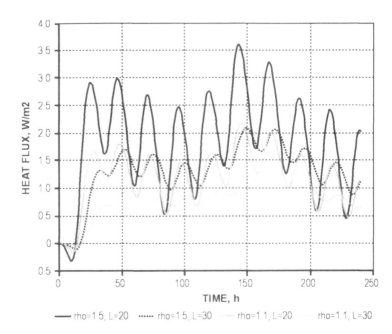

Fig. 5.21   Effect of the soil apparent density (rho) on the heat flux through the roof.

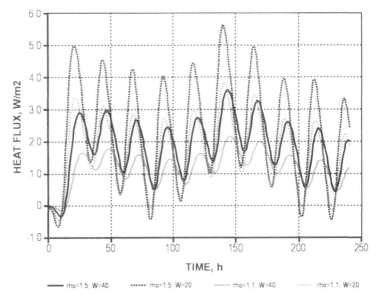

Fig. 5.22   Effect of the soil volumetric moisture content (*W*) on the heat flux through the roof.

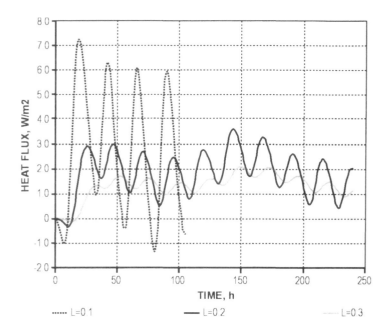

Fig. 5.23   Effect of the thickness of the soil layer (*L*) on the heat flux through the roof.

Finally, the thermal diffusivity of a soil, which affects both the speed at which a temperature pulse travels through the soil and the depth to which it may penetrate, is also affected by moisture: the thermal diffusivity of a dry soil will increase when water is added to it, reaching a maximum at about 20% soil moisture, but then begins to decline (Oke, 1987). In soils with high diffusivities, energy received at the surface warms a relatively thick layer, so changes in surface temperature are relatively small. Dry soils and most paving materials have lower diffusivities than moist soil, and consequently exhibit higher daytime temperatures and lower night-time temperatures. These properties are illustrated in Fig. 5.22. For the given soil densities, an increase in the soil moisture content results not only in a substantial reduction in the heat flux through the roof, but also in much smaller daily fluctuations.

The *thickness* of the soil substrate has a two-fold effect on the thermal performance of the roof (Fig. 5.23). First, the addition of material, in effect, increases the thermal insulation of the roof, thus reducing the rate at which heat is conducted through it; and second, since the addition of soil also increases the heat capacity of the roof, it reduces temperature fluctuations at the bottom of the soil layer. The heat flux through the roof over a period when environmental conditions fluctuate is therefore determined to a great degree by the extent to which the soil acts as a buffer: at a depth of about 80 cm (depending on the properties of the soil), temperature remains constant on a daily cycle (Oke, 1987).

Plant Canopy

Plant canopies are inherently complex, and any attempt to model them entails simplification and reduction of a number of numerical parameters. The following parameters were considered:

The *Leaf Area Index* (LAI), is the ratio of leaf area to ground surface area. A high value of LAI indicates a dense foliage. Foliage density affects the penetration of solar energy to the soil surface which, in turn, affects daytime surface temperatures (Fig. 5.24). The long-wave radiative heat loss from the surface is also affected, being reduced increasing with density of foliage, so night-time surface temperatures will tend to be higher with a dense foliage. Dense foliage will tend to reduce air speed in the canopy, allowing the development of a distinct microclimate; and a large surface area will result in greater transpiration (depending on plant species).

The proportion of incoming solar radiation that reaches the ground is affected not only by leaf density but also by leaf size, shape, colour and arrangement. Solar transmission declines exponentially with increasing values of the LAI and a *coefficient of extinction* representing the combined effect of these factors. High values of this coefficient indicate that most of the radiation is intercepted and absorbed by the leaf canopy (Fig. 5.25).

Transpiration (release of water vapour from leaves) occurs in the presence of sunlight, when photosynthesis takes place and stomata are open. Environmental conditions and the availability of water affect the rate of transpiration. However, different species vary in their propensity to transpire water. The contribution of transpiration from leaves to the creation of cooler and more humid conditions near plants has been one of the major reasons for the popularity of vegetation as a modifier of the local microclimate. The sensitivity study showed that the effect of transpiration on heat flux through a planted roof is, in fact, quite small. This is because transpiration affects heat flux only indirectly; it tends to cool the air in the plant

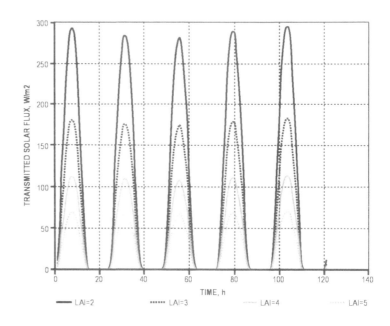

Fig. 5.24 Effect of the leaf area index on solar transmittance of the canopy.

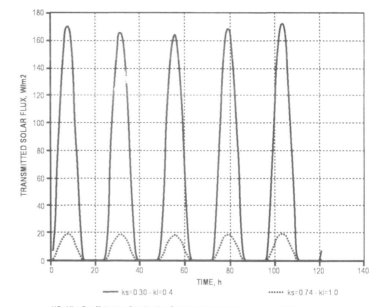

KS, KL: Coefficients of extinction for shortwave and longwave radiation

Fig. 5.25 Effect of the extinction coefficients on canopy solar transmission.

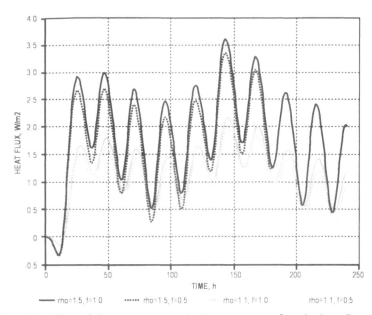

Fig. 5.26 **Effect of the canopy transpiration parameter *f* on the heat flux through the roof.**

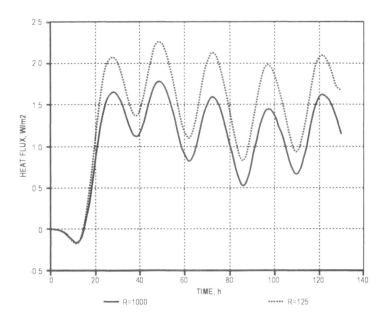

Fig. 5.27 **Effect of the parameter R on the heat flux through the roof.**

canopy, in turn lowering the surface temperature of the soil. However, this effect is very limited, since the reduction in air temperature is small and other factors such as radiant exchange and evaporation of soil moisture may have a dominant part in heat exchange at the surface of the soil (Fig. 5.26). The model introduced an artificial measure, *f*, which compares the tendency of a given plant to transpire water to that of tomatoes, which have been studied extensively.

All else being equal, *canopy height* determines the volume of air in which the energy exchanges occur. A low, dense canopy will tend to restrict air movement through it, magnifying the effects of evapotranspiration and heat and vapour exchange with the underlying soil.

Air Motion Through the Canopy

The rate at which the canopy exchanges air with the surrounding environment has a great effect on the capacity of the soil–canopy system to maintain conditions that are substantially different. A high air exchange rate (R=1000 air changes per hour) will tend to neutralise differences in air temperature and humidity, resulting in conditions that approximate ambient air. A low air exchange rate (R=125 air changes per hour), which represents an air velocity of about 0.03 m/s (a typical value for greenhouses without forced ventilation), permits the formation of a distinct microclimate in the plant canopy.

The effect of increasing the air exchange rate on heat flux through the roof depends on the properties of the roof. If there is an excess of water in the soil, so evaporation at the surface is not limited and, if the roof is well shaded by a dense leaf canopy, increasing air movement will tend to reduce heat gains (Fig. 5.27). This is due mainly to the increased evaporative cooling occurring at the surface of the soil (Theodosiou, 2003). If soil moisture content is low, the rate of evaporation becomes lower than the potential rate. Under these conditions, the capacity to maintain a distinct microclimate in the plant canopy is restricted, and it will become impossible to maintain a water vapour pressure that is higher than the surroundings in the canopy (Fig. 5.28). Additionally, if ambient air temperature is warmer than the interior temperature of the building, increasing the air exchange rate may result in higher convective gains at the soil surface.

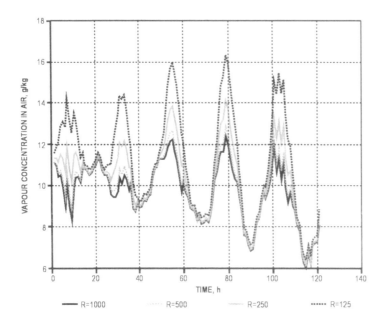

Fig. 5.28   Effect of the parameter R on the water vapour pressure in the air within the canopy.

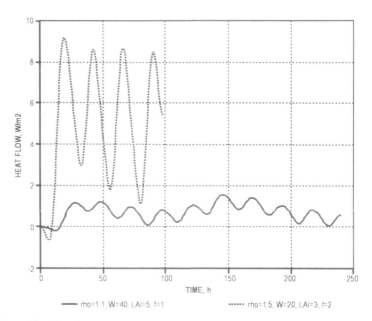

Fig. 5.29   Heat flux through two roofs representing extreme conditions for cooling.

The combined effect of a number of design parameters on heat flux through a planted roof is illustrated in Fig. 5.29. A poorly designed roof with a combination of high soil density, low moisture content, intermediate leaf canopy and little transpiration from leaves may subject the building interior to a heat load that is several times larger than a well-designed roof with a light soil, sufficient moisture and dense leaf canopy. It should be emphasised, though, that the studies presented above indicate that even the best planted roof cannot actually cool a building. It can, however, reduce external heat loads very substantially. This is because heat will be lost through the roof if the temperature of the interface between the structure of the roof and the soil above it is lower than the interior air temperature. The surface temperature of planted roofs during the warmest daytime hours may, in fact, be substantially lower than that of exposed concrete or conventional roof tiles. However, the combined effect of the plant cover, soil and moisture do not result in a reduction of the soil temperature below ambient air temperature.

Designing a planted roof that reduces heat flux through the element in summer requires:

- *plant selection*: select plants with extensive foliage and/or with mainly horizontal leaves; these result in low solar transmission and provide better shading;
- *soil selection*: select light soils which reduce thermal conductivity, as well as weight.

The beneficial thermal effects of planted roofs in warm climates are the result of two main mechanisms: protection from solar radiation and evaporation of soil moisture. It follows therefore that the greatest effect of planted roofs will be felt in locations where the climate is sunny and dry. Where the climate is warm but overcast and humid for much of the time, the gains from planted roofs will be much smaller. Consequently, when the outdoor environment is warmer than the indoor spaces, the effect of planted roofs is similar to that of thermal insulation. When the outdoor environment is colder than the building interior, however, the thermal attributes of the planted roof are of no benefit. The performance of planted roofs in winter depends on the ability of the soil to provide sufficient thermal insulation. Achieving sufficient thermal insulation to comply with standards in most European countries would require an inordinate amount of soil, imposing unacceptable structural loads on the supporting structure.

A simulation of the relative benefits of a plant layer on roofs with varying amounts of thermal insulation (Niachou *et al.* 2001) showed that in a climate such as that of Athens, Greece, substantial energy savings may be made only if the roof has little or no insulation to begin with. For a well-insulated roof (*U*-value of 0.4 W/m^{2}K or lower), seasonal energy savings from adding a roof garden were only 2%, whether or not night ventilation was employed as an additional cooling strategy.

References

Köhler, M., Schmidt, M., Grimme, F.W., Laar, M., de Assunção Paiva, V.L. and Tavares, S. (2002) Green roofs in temperate climates and in the hot-humid tropics. *Environmental Management and Health*, 13(4): 382–391.

McPherson, G., Herrington, L. and Heisler, G. (1988) Impacts of vegetation on residential heating and cooling. *Energy and Buildings* 12: 41–51.

McPherson, G., Simpson, J. and Livingston, M. (1989) Effects of three landscape treatments on residential energy and water use in Tucson, Arizona. *Energy and Buildings*, 13: 127–138.

Niachou, N., Papakonstantinou, K., Santamouris, M., Tsangrassoulis, A. and Mihalakakou, G. (2001) Analysis of the green roof thermal properties and investigation of its energy performance. *Energy and Buildings*, 33: 719–729.

Oke, T. (1987) *Boundary Layer Climates*. 2nd edn. Routledge, London.

Palomo Del Barrio, E. (1998) Analysis of the cooling potential of green roofs in buildings. *Energy and Buildings*, 27: 179–193.

Theodosiou, T. (2003) Summer period analysis of the performance of a planted roof as a passive cooling technique. *Energy and Buildings*, 35: 909–917.

Wong, N., Chen, Y., Ong, C., Cheong, D. and Sia, A. (2003). Investigation of thermal benefits of rooftop garden in the tropical environment. *Building and Environment*, 38: 261–270.

6 APPLICABILITY & PERFORMANCE DATA

INTRODUCTION

This chapter provides comparative performance data for a selection of roof cooling techniques discussed in this handbook. The performance data give a measure of cooling energy demand and possible energy savings on air-conditioning for mechanically cooled buildings, and likely improvements in indoor thermal conditions for non-air-conditioned buildings. These were obtained from simulations performed on different building specifications with and without the roof cooling applications. The simulations were performed with the RSPT software, which is on the CD included with this publication. The data in this chapter give a first indication of applicability and comparative performance of roof cooling techniques in different climatic conditions. With these as a starting point readers can then run the RSPT software on building configurations and roof cooling specifications of their own for locations of their choice. Use of the software is described in Appendix B.

ENERGY DEMAND FOR SPACE HEATING AND COOLING

The maps of parts of Europe in Figs 6.1 and 6.2 provide measures of comparative space heating and cooling energy demand for residential and office buildings. The coloured areas represent different bands of energy demand in kWh per square metre of building floor area. These are a measure of the energy needed to maintain the indoor temperature to setpoints of 20°C in the heating season and 25°C in the cooling season. Geographically, the demand for cooling peaks in the south of the continent where summer sunshine and outdoor temperatures are highest. In terms of building types, the demand for cooling increases with density of occupation and the rate of internal heat generation. Over the last ten years the demand for air-conditioning equipment has been growing at a very fast rate across Europe. With climate change predicted to continue increasing building cooling loads in the near future, reducing dependency on air-conditioning acquires further importance.

CLIMATIC APPLICABILITY

The maps in Figs 6.1 and 6.2 highlight considerable differences in space heating and cooling demands that result from climatic variations due to geographic location. The ambient dry-bulb temperature is quite interesting in this respect. It is a source of heat gain, and thus of cooling load at times, whilst being a heat sink in other periods of the day and year. For example, if we were to take a dry-bulb

 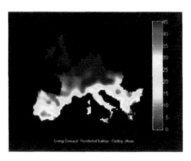

Fig. 6.1 Indicative space heating (left) and cooling (right) demand in kWh per square metre floor area for a residential building in different parts of Europe.

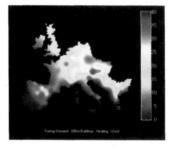 

Fig. 6.2 Indicative space heating (left) and cooling (right) demand in kWh per square metre floor area for environmentally poor office buildings in different parts of Europe.

temperature of 25°C as the base temperature for indoor thermal comfort in summer, the ambient air could impose a cooling load whenever its temperature exceeded this value. We can obtain a relative measure of this cooling load by counting the number of days on which the outdoor air temperature might exceed this base temperature for cooling, and the magnitude of this excess. This temperature–time product is a measure of *cooling degree-days* for a given location. For example, the following are cooling degree-days to base 25°C (CDD_{25}) for four European cities over the period of June–September inclusive:

Athens, Greece   319 CDD_{25}
Seville, Spain   316 CDD_{25}
Madrid, Spain   125 CDD_{25}
Paris, France   23 CDD_{25}

(a) Dry-bulb degree days
(24 hour day)

(b) Wet-bulb depression
(0600–2200 hours)

(c) Wet-bulb depression
(2200–0600 hours)

(d) Sky temperature depression
(2200–0600 hours)

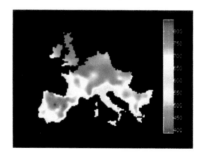

Fig. 6.3   Maps of summertime degree-days (to base 25°C) and temperature depressions (to base outdoor air temperature) in regions of Europe.

In southern European locations such as Athens and Seville, as well as in other parts of the world with hot summers, the ambient dry-bulb temperature may rise above 25°C for much of the daytime period in the cooling season*. In such locations this parameter will be a source of heat gain at these times, but may act as a heat sink providing a cooling potential at night-time when its value normally drops below 25°C. In some other regions, on the other hand, the ambient dry-bulb temperature may act as a heat sink even during daytime. This potential for convective cooling is shown by the map in Fig. 6.3(a), which is drawn by plotting the number of degree-days on which the ambient dry-bulb temperature is *lower* than 25°C in the cooling season. As might be expected, this potential is much higher in the north than in the southern regions of the continent.

When the ambient dry-bulb temperature is not low enough to act as a heat sink, there are two other climatic parameters which may do so. These are the ambient wet-bulb temperature and the sky temperature (see Chapter 1 for definitions). Table 6.1 gives the cooling season average and mean maximum and mean minimum values of these parameters for Seville, Spain. The values of the dry-bulb temperature indicate a very good potential for cooling by night ventilation owing to the substantial drop in the ambient air temperature. A large incentive for high thermal inertia is indicated by the very high mean outdoor maxima and the proximity of the daily average to comfort values. The difference between the dry-bulb and wet-bulb temperatures suggests good potential for evaporative cooling, especially during daytime, owing to the very high dry-bulb values.

TABLE 6.1   MEAN DRY-BULB (T_{db}) WET-BULB (T_{wb}) AND SKY (T_{sky}) TEMPERATURES (°C), SEVILLE

	T_{db}	T_{wb}	T_{sky}
Mean max	41.1	22.3	26.1
Mean	25.5	17.8	11.5
Mean min	13.9	12.3	0.5

Similarly the difference between dry-bulb and sky temperatures indicates the likelihood of good potential from radiative cooling. The sums over time of those temperature differences, respectively the *wet-bulb temperature depression* and the *sky temperature depression,* are illustrated in Fig. 6.3(b), (c), (d). The higher the numerical values of these temperature depressions, the higher will be the resulting cooling potential. The wet-bulb depression appears on two separate maps, one for the daytime hours (taken as the period from 0600 to 2200 hours) and one for night-time (2200–0600 hours). It can be seen that for southern Europe, the evaporative cooling potential is quite high, both during daytime and at night. Overall, in southern Europe, the evaporative cooling potential (both daytime and night) is in the range of 500–1000 degree-days, similar to that available from radiative cooling at night. Convective cooling of the building structure is mainly available at night. For northern Europe the main cooling potentials are convective and radiative.

* In all of the calculations presented in this chapter the cooling season is the period between June and September inclusive (a total of 122 days). In practice the cooling season may be longer or shorter than this, depending on the thermal and operational characteristics of buildings and the cooling setpoint temperature assumed.

REFERENCE BUILDING TYPES AND CONSTRUCTIONS

The cooling techniques for which performance data are given in this chapter were considered on single-storey and multi-storey residential and office buildings. Values of potential energy savings and thermal comfort improvements were obtained by comparison with reference buildings of the same specification except for the roof cooling devices. The reference buildings were assumed to be of conventional construction compliant with local Building Regulations. For the purpose of providing numerical data in this chapter the reference buildings were modelled on typical southern European construction practice and their thermal characteristics were based on the new Spanish Code on thermal insulation for energy conservation in buildings. For each roof cooling configuration, a base case and a number of variants were modelled and assessed.

The base-case reference residential building was a single-storey detached dwelling of 100 m^2 occupied floor area. External walls were assumed to be of cavity construction with hollow and solid brick leaves and 40 mm mineral wool insulation in cavity. This had an overall thermal transmittance (U-value) of 0.66 W/m^{2}K. External windows were assumed to be fitted with double glazing with a total area of 10 m^2 distributed evenly between north, east and south elevations. Two different reference roof constructions were associated with this building model. One consisted of the typical roof construction likely to be adopted in conventional practice. This comprised a ceramic layer of 200 mm (with a thermal capacity of 220 kJ/m^{2}K) and a 50 mm concrete slab (of thermal capacity 105 kJ/m^{2}K) over a plastered ceiling. This roof was insulated with a 40 mm thick polystyrene slab above the structure. The overall thermal transmittance was calculated at 0.58 W/m^{2}K. This typically heavyweight roof is identified as **HI** in the following sections of this chapter. The second reference roof construction was of lightweight construction consisting solely of a structural steel deck. For the purpose of comparative performance assessments the thermal insulation of this roof construction was considered in two variants: (1) with a 40 mm layer of mineral wool insulation above the steel deck, (2) with 62 mm mineral wool insulation above the steel deck. These roof constructions are identified with the codes **LI** and **LI+**, respectively. The resulting thermal transmittance values were calculated as 0.79 W/m^{2}K and 0.54 W/m^{2}K. A further roof configuration was modelled. This simply consists of a steel deck without any fixed thermal insulation layer. This is identified with the code **L** and has a calculated U-value of

5.88 W/m^{2}K. It is not a reference roof since it would not meet Building Regulations in this form. It is, however, a very important base case for roof pond applications.

On this single-storey dwelling all roof configurations were assumed to have a surface area equal to the dwelling floor area. The reference dwelling was assumed to have four occupants with metabolic gains of 45.4 W per person latent and 71.8 W per person sensible. Internal gains from lights were taken as 4.4 W/m^2 and for equipment 4.4 W/m^2. For the air-conditioned variant the AC system was assumed to operate continuously to maintain indoor temperature to a thermostat setpoint of 25°C. Two-storey dwellings were also considered for the performance assessments included here. These have the same envelope specifications as those described above.

ROOF COOLING VARIANTS

Roof Pond Systems

On single-storey dwellings, roof ponds were assumed to cover the entire roof area. The pond water was assumed to be contained within the roof parapet over the roof constructions corresponding to the L, LI, LI+ and HI specifications introduced above. A movable insulated cover was configured with each of these specifications in addition to the fixed layer of insulation they were assigned, see above. The roof configuration identified by the letter L can be assumed to satisfy its regional thermal insulation standards by maintaining this insulated cover in place throughout the winter heating season. This would achieve a U-value of 0.37 W/m^{2}K which should meet Building Regulations requirements in most parts of southern Europe. For the other three roof variants the insulated cover lowers their U-values to around 0.25 W/m^{2}K when in place.

The system parameters that defined the roof pond applications were: the water depth in the pond; the properties and operation of the spraying system; the water circulation rate; the pond cover properties and operation. The following values were adopted for base case and parametric variations:

- **pond water depth**
  base case: 0.3 m
  performance assessed for range: 0.1–0.5 m
- **water circulation**
  base case: 1 pond volume per hour

performance assessed for range: 0.0–3.0 water changes per hour
- **spray droplet radius**
  base case: 1.0 mm
  performance assessed for range: 0.5–3.0 mm
- **jet height**
  base case: 1.3 m
  performance assessed for range: 0.5–1.5 m.
- **pond cover properties**
  solar reflectivity: 0.80
  thermal conductivity: 0.036 W/mK
  thickness: 100 mm
  no air gap between cover and pond water surface.

Three alternative operational schedules were considered for the pond cover and spraying system:

- **C-N**: continuous spraying, no pond cover;
- **N-D**: spraying at night only, pond covered daytime;
- **C-D**: continuous spraying, pond covered daytime.

A number of other permutations were also considered (no spraying, continuous use of cover) and have been incorporated in the RSPT software.

Roof Pond with Cooling Panels

On multi-storey buildings the roof pond was considered to be coupled to cooling panels fitted on the building's intermediate floors. The cooling panels had the following base case specification:

- **concrete slab** 60 mm thick attached to ceiling
- **parallel pipes** 18 mm in diameter
  base case: spacing at 200 mm centres
  performance assessed for range spacing: 100–300 mm centres
- **water flow** from roof pond
  base case flow: 12 kg/s
  performance assessed for range: 6–18 kg/s
- water circulated through one or more cooling panels depending on number of floors served.

Three operational variants were considered:

- C: water running through the panels continuously
- N: water running at night only
- D: water running daytime only.

In the results shown in this chapter and in Appendix A the cooling panels were coupled to a roof pond with the base case specification as listed above.

Cooling Radiators

Air- and water-based radiator parameters were varied in terms of dimensions, mass flows and operational schedules. Water radiators were studied in combination with roof ponds or with cooling panels.

The following values were used for water radiator base case specification and variants:

- **length**
  base case: 6.0 m
  performance assessed for range: 2.0–10.0 m
- **mass flow rate**
  base case: 0.03 kg/s
  performance assessed for range: 0.01–0.1 kg/s
- **pipes per metre**
  base case: 50
  performance assessed for range: 10–50.

The following values were used for air radiator base case specification and variants:

- **length**
  base case: 6.0 m
  performance assessed for range: 2.0–10.0 m
- **air channel thickness**
  base case: 30 mm
  performance assessed for range: 10–50 mm
- **air flow rate**
  base case: 0.03 kg/s
  performance assessed for range: 0.03–0.1 kg/s

The following operational schedules were considered:

- N: night-time operation only;
- D: daytime only;
- C: continuous operation.

Planted Roofs

Following the findings reported in Chapter 5 no further calculations were performed on planted roofs and the model was not implemented in the RSPT software.

APPLICABILITY AND PERFORMANCE DATA

Introduction

In order to provide unbiased comparative data, each roof cooling configuration was assessed by comparison with a reference case that excludes the roof cooling system but is otherwise based on the same building specification and general operational conditions. The maps, graphs and tables that follow provide comparative performance data on the following:

- *energy demand* for space cooling in kWh based on continuous occupancy and system operation to 25°C over the four months June to September inclusive;
- *energy savings* on air-conditioning as the difference in energy demand between reference case and building configuration with roof cooling system, expressed in kWh or as a percentage.

For free-running (non-air-conditioned) buildings, comparative performance data are given on the following:

- total *number of hours* in the cooling season (defined as above) with mean indoor temperature above 25°C and percentage of the daily occupancy period with indoor temperature above this threshold;
- *peak indoor temperature* in the cooling period.

The cooling energy demand figures allow comparisons to be drawn between alternative roof cooling systems. The thermal comfort data (peak indoor temperatures and hours above 25°C or other indoor temperature threshold) can serve two purposes. They provide an indication of the likely improvement in thermal comfort that can be achieved in free-running buildings. They can also help decide whether or not to install an air-conditioning system.

Summer Performance Data for Seville

As an example the tables and graphs on the following pages summarise performance predictions for single-storey residential buildings in the climatic conditions of Seville (Latitude 37°24'N), southern Spain. In the peak summer months of July and August, Seville experiences more than 11 hours of daily sunshine and the average ambient air temperature peaks above 36°C, with mean daily averages around 28°C. The calculations show that without a conventional air-conditioning appliance and no roof cooling application in place, the indoor temperature in the reference dwellings would rise above 25°C for a total of over 1700 hours (this is equivalent to 14 hours out of every 24-hour period) reaching peaks between 32.2°C and 36.8°C. Clearly this level of overheating encourages occupants to install air conditioning, thus perpetuating the recent trend. The cooling energy demand in order to maintain the reference buildings to 25°C on a continuous regime was calculated in the range of 1700–2400 kWh (17–24 kWh per square metre floor area). Of the three reference roof constructions the one with the heavyweight roof (HI) achieved the lower energy demand when air-conditioned, and resulted in fewer hours above 25°C and lower peak temperatures when free-running. This is because of the heat storage capacity this configuration provides in the roof structure compared with the lightweight variants. The LI+ variant, which has a thicker layer of roof insulation than the LI variant, is shown to have a lower cooling requirement than the latter. The L variant was assumed to have the same reference as LI. These results are shown in Table 6.2 together with base case and best roof pond configurations.

The base case roof pond application has the following specification:

- pond water depth: 0.3 m
- water circulation: 1 volume per hour
- spray droplet radius: 1.0 mm
- jet height: 1.3 m
- CS-XC or DC: continuous spraying without pond cover or with daytime cover.

The best performance for the above specification is obtained with a daytime cover on the roof pond. The results summarised in Table 6.2 show that the base case roof pond could lead to energy savings in the region of 44–82%, without daytime cover, compared with the reference cases. With the insulated cover in place the savings

TABLE 6.2   SUMMARY PERFORMANCE DATA FOR ROOF PONDS, SEVILLE

ROOF CONSTRUCTION		COOLING ENERGY DEMAND					COMFORT					
		REF	BASE CASE		WITH COVER		REF		BASE CASE		WITH COVER	
		kWh	kWh	%	kWh	%	Hours>25°C	Peak	Hours	Peak	Hours	Peak
L	steel deck, no fixed insulation	2394	440	82	72	97	1723	36.8	250	28.0	7	25.3
LI	steel deck, 40 mm mineral wool	2394	1064	56	905	62	1723	36.8	820	30.7	702	30.0
LI+	steel deck, 62 mm mineral wool	2084	1158	44	1025	51	1706	36.1	902	31.1	819	30.6
HI	250 mm masonry, 40 mm polystyrene	1718	896	48	802	53	1765	32.2	660	28.5	570	28.3

increase to 51–97%. The highest savings result from roof variant **L**, owing to the better coupling that this configuration has with the building interior. Adopting this configuration appears to make air-conditioning redundant. On the other variants the thermal coupling between the roof pond and the occupied spaces appears to be inhibited by the mediation of the thermal insulation layers required to satisfy the Building Regulations.

Tables 6.3 and 6.4 and related graphs overleaf summarise the performance results for the full range of parameters analysed for single-storey dwellings in the climatic conditions of Seville. These may be used at the design stage in conjunction with the RSPT software to review design options. The areas marked in grey identify the base case values of parameters. Generally, these represent the resulting quasi-optimum values between energy and water consumption. In Table 6.3 the four base case roof construction types are identified on the left margin with the reference energy demand for summer cooling to an indoor temperature of 25°C. For each of these roof types the table identifies the likely percentage energy savings that can be achieved by varying roof pond design parameters and operational schedules. It can be seen that further savings can be obtained beyond those achieved by the base case roof pond parameters. However, these will tend to increase water consumption. In Tables 6.2 and 6.4 the L variant can be seen again to lead to best results in terms of thermal comfort for free-running buildings. Of the other roof types, HI has the best results in terms of peak temperature and hours above the set temperature. Figure 6.4 plots the evolution of indoor temperatures over seven consecutive warm days, further highlighting these results. Note how the roof

ponds reduce both the mean value and the fluctuations of the indoor temperature lowering it toward the outdoor wet-bulb. It is notable that for the selected period all of the roof cooling variants lead to mean daily temperatures that are lower than the air-conditioning setpoint.

Table 6.5 compares summary performance data for Seville, Athens, Madrid and Paris. These locations were selected as representing most of the spread of summer climatic conditions around Europe. Seville displays the largest fluctuation between daytime and night-time ambient temperatures. This is accompanied by the highest   wet-bulb and sky temperature depressions, thus providing the largest potential for roof cooling techniques. Athens has the highest cooling loads. This is due to the smaller fluctuation between daytime and night-time ambient temperatures, with both remaining at high values. The wet-bulb and sky temperature depressions are lower than in Seville, resulting in a smaller contribution from roof cooling techniques. For Madrid, the cooling loads are much lower than for Athens and Seville, but the potential for meeting them with roof ponds appears to be very good. The results show that roof ponds could achieve 100% savings and indoor temperatures well within comfort for free-running buildings for any of the roof type configurations. The case of Paris is shown here to illustrate that as we move towards the north of Europe cooling loads are much reduced and, for well-designed residential buildings, there should be no need to consider air-conditioning. The roof pond results show that cool indoor temperatures can be achieved throughout the summer period despite peak outdoor temperatures above 32°C.

TABLE 6.3 COOLING ENERGY SAVINGS FOR RESIDENTIAL ROOF POND VARIANTS AND OPERATIONAL SCHEDULES

ROOF TYPE: L

REF COOLING -2394

Water in parapet
Metal deck

POND DEPTH (m)					WATER CHANGES PER HOUR				DROP RADIUS (mm)				JET HEIGHT (m)				SCHEDULES		
0.1	0.2	0.3	0.4	0.5	0.0	1.0	2.0	3.0	0.5	1.0	2.0	3.0	0.5	1.0	1.3	1.5	C-N	N-D	C-D
																	-81.6%	-96.9%	-97.0%
-50.5%	-72.2%	-81.6%	-86.2%	-89.0%															
					-32.9%	-81.6%	-88.6%	-91.3%											
									-89.0%	-81.6%	-67.0%	-57.3%							
													-73.7%	-79.7%	-81.6%	-82.6%			

ROOF TYPE: LI

REF COOLING -2394

Water in parapet
Waterproofing
Thermal insulation
Metal deck

POND DEPTH (m)					WATER CHANGES PER HOUR				DROP RADIUS (mm)				JET HEIGHT (m)				SCHEDULES		
0.1	0.2	0.3	0.4	0.5	0.0	1.0	2.0	3.0	0.5	1.0	2.0	3.0	0.5	1.0	1.3	1.5	C-N	N-D	C-D
																	-55.6%	-58.8%	-62.2%
-48.6%	-53.1%	-55.6%	-56.8%	-57.8%															
					-42.8%	-55.6%	-58.0%	59.2%											
									-58.1%	-55.6%	-51.0%	-48.5%							
													-52.9%	-54.6%	-55.6%	-55.7%			

ROOF TYPE: LI+

REF COOLING -2084

as LI and additional
thermal insulation

POND DEPTH (m)					WATER CHANGES PER HOUR				DROP RADIUS (mm)				JET HEIGHT (m)				SCHEDULES		
0.1	0.2	0.3	0.4	0.5	0.0	1.0	2.0	3.0	0.5	1.0	2.0	3.0	0.5	1.0	1.3	1.5	C-N	N-D	C-D
																	-44.4%	-50.8%	-50.1%
-38.7%	-42.5%	-44.4%	-45.6%	-46.5%															
					-34.1%	-44.4%	-46.8%	-47.4%											
									-46.7%	-44.4%	-40.9%	-38.6%							
													-42.6%	-44.1%	-44.4%	-44.7%			

ROOF TYPE: HI

REF COOLING -1718

Water in parapet
Waterproofing
Insulation
Masonry structure

POND DEPTH (m)					WATER CHANGES PER HOUR				DROP RADIUS (mm)				JET HEIGHT (m)				SCHEDULES		
0.1	0.2	0.3	0.4	0.5	0.0	1.0	2.0	3.0	0.5	1.0	2.0	3.0	0.5	1	1.3	1.5	C-N	N-D	C-D
																	-47.8%	-53.3%	-53.0%
-43.0%	-46.0%	-47.8%	-48.8%	-49.6%															
					-34.4%	-47.8%	-50.1%	-51.0%											
									-50.0%	-47.8%	-43.7%	-40.9%							
													-45.4%	-47.5%	-47.8%	-48.4%			

KEY

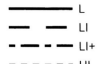

——————— L
— — — LI
— · — · — LI+
- - - - - HI

% COOLING ENERGY SAVINGS FOR ROOF POND VARIANTS

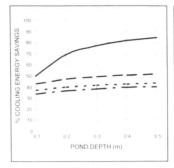

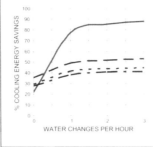

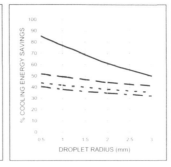

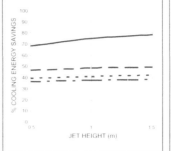

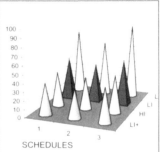

TABLE 6.4　PEAK INDOOR TEMPERATURES AND HOURS ABOVE 25°C FOR ROOF POND VARIANTS AND OPERATIONAL SCHEDULES

ROOF TYPE: L

POND DEPTH (m)					WATER CHANGES PER HOUR				DROP RADIUS (mm)				JET HEIGHT (m)				SCHEDULES		
0.1	0.2	0.3	0.4	0.5	0.0	1.0	2.0	3.0	0.5	1.0	2.0	3.0	0.5	1.0	1.3	1.5	C-N	N-D	C-D
																	28.0	25.3	25.3
30.7	29.0	28.0	27.4	27.0															
					31.2	28.0	27.1	26.7											
									27.0	28.0	29.3	30.0							
													28.8	28.3	28.0	27.9			

ROOF TYPE: LI

POND DEPTH (m)					WATER CHANGES PER HOUR				DROP RADIUS (mm)				JET HEIGHT (m)				SCHEDULES		
0.1	0.2	0.3	0.4	0.5	0.0	1.0	2.0	3.0	0.5	1.0	2.0	3.0	0.5	1.0	1.3	1.5	C-N	N-D	C-D
																	30.7	30.0	30.1
31.3	30.9	30.7	31.1	30.5															
					31.6	30.7	30.5	30.4											
									30.5	30.7	31.1	31.2							
													30.9	30.8	30.7	30.7			

ROOF TYPE: LI+

POND DEPTH (m)					WATER CHANGES PER HOUR				DROP RADIUS (mm)				JET HEIGHT (m)				SCHEDULES		
0.1	0.2	0.3	0.4	0.5	0.0	1.0	2.0	3.0	0.5	1.0	2.0	3.0	0.5	1.0	1.3	1.5	C-N	N-D	C-D
																	31.1	30.6	30.8
31.4	31.2	31.1	31.1	31.0															
					31.7	31.1	31.0	30.9											
									31.0	31.1	31.4	31.4							
													31.2	31.1	31.1	31.1			

ROOF TYPE: HI

POND DEPTH (m)					WATER CHANGES PER HOUR				DROP RADIUS (mm)				JET HEIGHT (m)				SCHEDULES		
0.1	0.2	0.3	0.4	0.5	0.0	1.0	2.0	3.0	0.5	1.0	2.0	3.0	0.5	1	1.3	1.5	C-N	N-D	C-D
																	28.5	28.3	28.3
28.7	28.6	28.5	28.5	28.5															
					29.2	28.5	28.4	28.4											
									28.5	28.5	28.7	28.8							
													28.7	28.5	28.5	28.5			

SEVILLE

ROOF TYPE: L
REF PEAK: **36.8°C**
Water in bags
Metal deck

ROOF TYPE: LI
REF PEAK: **36.8°C**
Water exposed
Waterproofing
Thermal insulation
Metal deck

ROOF TYPE: LI+
REF PEAK: **36.1°C**
as **LI** and additional
thermal insulation

ROOF TYPE: HI
REF PEAK: **32.2°C**
Water exposed
Waterproofing
Insulation
Masonry roof struc-

KEY

—— L
— — LI
-·-·- LI+
- - - - HI

NUMBER OF HOURS INDOOR TEMPERATURE >25°C

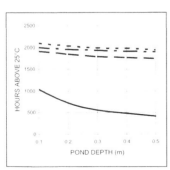

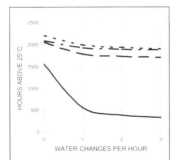

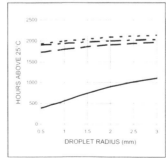

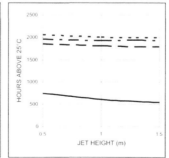

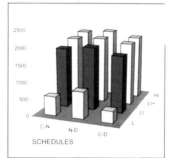

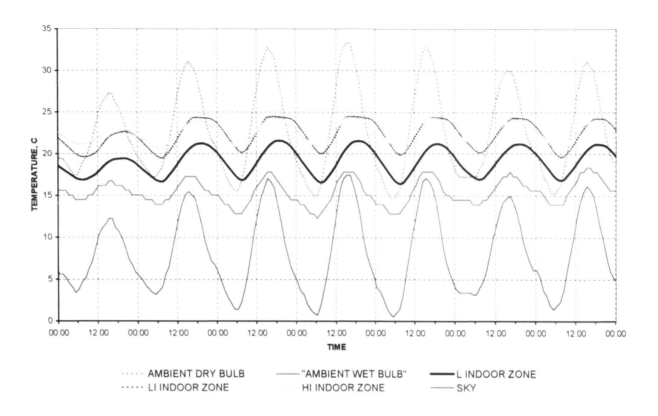

Fig. 6.4 Evolution of predicted indoor temperature for three of the roof types coupled to roof ponds for climatic conditions of Seville.

Figure 6.4 shows the evolution of predicted indoor temperatures over a period of a week in summer for the climatic conditions of Seville. The cases selected are the insulated lightweight LI and heavyweight HI roof constructions and the uninsulated L roof construction. The roof ponds are base case with the cover in place during daytime but the spraying system was assumed to be off throughout. The roof ponds coupled to the LI and HI configurations results in very similar indoor temperatures that stay below 25°C thus providing even cooler conditions than a 24-hour air-conditioning set at 25°C in this period. The roof pond coupled to the L configuration yields even lower temperatures. These are plotted against the background of the ambient dry-bulb, wet-bulb and sky temperatures.

Table 6.5 provides a summary overview of differences in cooling load and roof pond performance for four climatically distinct European locations. The maps illustrated in the following pages give a more detailed picture of variations in cooling loads and roof cooling potential across Europe. As with the tables, the maps plot data for both air-conditioned and free-running cases. The two maps for the former give the energy savings from roof ponds in absolute value (expressed in kWh) and as a percentage. The free-running maps give the total number of hours predicted with indoor temperatures above 25°C and the percentage of the 24-hour occupancy period represented by these hours. The maps were produced with the RSPT software that is included with this publication.

TABLE 6.5 **COMPARATIVE PERFORMANCE DATA FOR FOUR EUROPEAN LOCATIONS**

ROOF CONSTRUCTION	SEVILLE				ATHENS				MADRID				PARIS			
	REF kWh	PEAK °C	SAVE %	PEAK °C	REF kWh	PEAK °C	SAVE %	PEAK °C	REF kWh	PEAK °C	SAVE %	PEAK °C	REF kWh	PEAK °C	SAVE %	PEAK °C
L steel deck, no fixed insulation	2394	36.8	97	25.3	3278	36.9	89	28.0	947	34.3	100	21.9	159	32.7	100	21.7
HI masonry, 40 mm insulation	1718	32.2	53	28.3	2805	33.8	44	30.6	436	30.1	77	26.1	20	27.1	100	22.7

Water Consumption

Water is consumed by evaporation from open roof ponds and through spraying. The base case roof ponds analysed for this section (100 m² floor area and 0.3 m depth) contained some 30 m³ of water that was circulated once per hour. On different variants depth was varied up to 0.5 m and water circulation up to three changes per hour. It was estimated that the resulting water consumption for these configurations could vary between 0.42 and 1.14 m³ per day in the climatic conditions of Athens, 0.17 and 1.05 m³ per day in the climatic conditions of Seville, and 0.18 and 0.96 m³ per day in the climatic conditions of Madrid. Water consumption can be reduced by containing the pond water in plastic bags (see Chapter 3). It can be further reduced by closer control of the use of the spraying system. Finally, it can eliminated by not using a spraying system and relying solely on longwave radiation for cooling.

Winter Space Heating Energy Demand

Sample calculations showed that for conventional roof constructions such as those assumed for the performance data given here it may be desirable to drain the roof ponds in winter and have the roof operate as a conventionally insulated element. The thermal insulation level for the roof can be specified according to local Building Regulations.

ROOF PONDS   APPLICABILITY MAPS FOR BEST CASE ROOF PONDS ON SINGLE-STOREY RESIDENTIAL BUILDINGS

Best Result
**Residential
Single-storey**

Pond depth: 0.3 m
Spray height: 1.3 m
Droplet radius: 1 mm
Water changes per
hour (wch): 1.0
Cover: daytime
Spraying: Continuous

The roof pond variants
can achieve energy sav-
ings of 50–100% on con-
ventional air-conditioning
based on a thermostat
setting of 25°C and 24-
hour operation.

WITH AIR-CONDITIONING

COOLING ENERGY SAVINGS, KWH

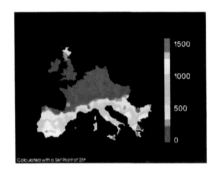

% ENERGY SAVINGS BY ROOF POND

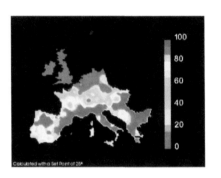

FREE-RUNNING

NUMBER OF HOURS >25°C

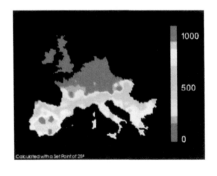

HOURS >25°C
AS % OF OCCUPANCY PERIOD

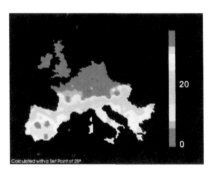

APPLICABILITY MAPS FOR BEST CASE ROOF POND AND COOLING PANEL FOR TWO-STOREY RESIDENTIAL BUILDING

WITH AIR-CONDITIONING

COOLING ENERGY SAVINGS, KWH

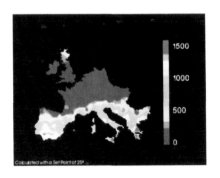

% ENERGY SAVINGS

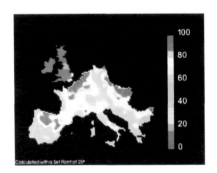

The best case roof pond with cooling panel variants can achieve energy savings of 30–100% in different European locations compared with conventional air-conditioning with a thermostat setting of 25°C and 24-hour operation.

Pond depth: 0.3 m
Spray height: 1.3 m
Droplet radius: 1 mm
wch (night-time): 1.0
Cover: daytime

Cooling panel: 10 m
Pipes: 18 mm @100 mm
Water flow: 12 kg/s

Two-storey
Residential

200 m² floor area

FREE-RUNNING

NUMBER OF HOURS >25°C

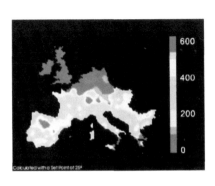

HOURS >25°C
AS % OF OCCUPANCY PERIOD

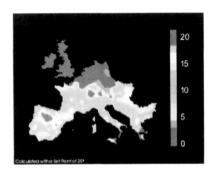

The best variants achieve reductions in peak temperature of 7–9K.

COOLING PANEL

APPLICABILITY MAPS FOR BEST CASE COOLING PANEL ON TWO-STOREY RESIDENTIAL BUILDINGS

Best Result
Residential

Two-storey dwelling

Cooling panel fed
fresh water from
spring or river.

WITH AIR-CONDITIONING

COOLING ENERGY SAVINGS, KWH

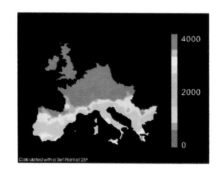

% ENERGY SAVINGS BY ROOF POND

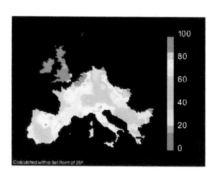

FREE-RUNNING

NUMBER OF HOURS >25°C

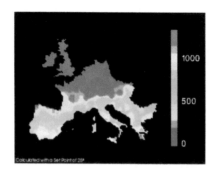

HOURS >25°C
AS % OF OCCUPANCY PERIOD

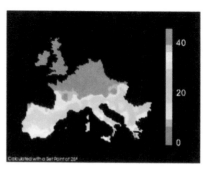

APPLICABILITY MAPS FOR BEST CASE WATER RADIATOR & COOLING PANEL FOR 2-STOREY RESIDENTIAL BUILDINGS

**WATER
RADIATOR &
COOLING PANEL**

WITH AIR-CONDITIONING

COOLING ENERGY SAVINGS, KWH

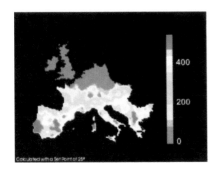

% ENERGY SAVINGS

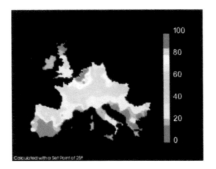

Cooling panel: 10 m
Pipes:18 mm @100 mm
water flow: 12 kg/s

water radiator
length 4 metres
width 5 metres
pipes per metre 15
distance between pipes
0.058

Two-storey
Residential

FREE-RUNNING

NUMBER OF HOURS >25°C

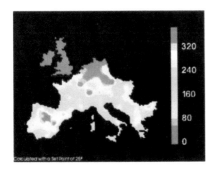

HOURS >25°C
AS % OF OCCUPANCY PERIOD

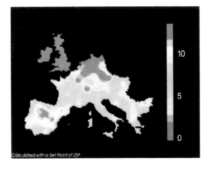

CLIMATE INDEX

ROOF PONDS

Best Case
with continuous
spraying and day-
time cover

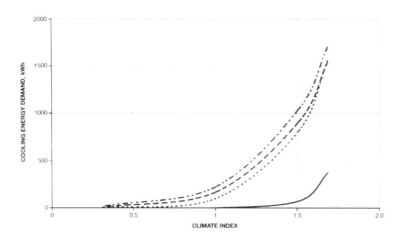

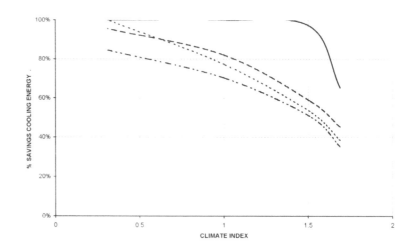

KEY

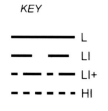

L
LI
LI+
HI

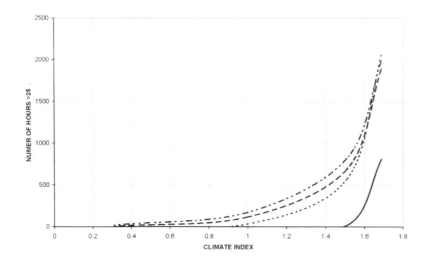

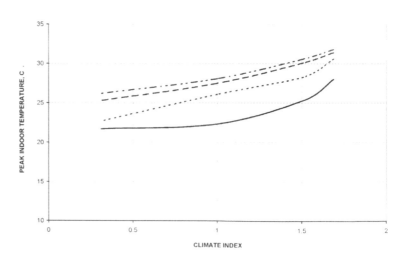

Calculation and Use of the Climate Index (CI)

Step 1 Calculation of CI from degree-days and global solar radiation

$$CI = a \cdot RAD + b \cdot DD + c \cdot RAD \cdot DD + d \cdot (RAD)^2 + e \cdot (DD)^2 + f$$

where,

CI is climate index;
RAD is mean value of monthly global solar radiation on the horizontal for selected location and period June–September in kWh/m²;
DD is mean value of monthly cooling degree-days to base 20 (ambient air temperature minus 20) for June to September;

and:

a	b	c	d	e	f
$3.724 \cdot 10^{-3}$	$1.409 \cdot 10^{-2}$	$-1.869 \cdot 10^{-5}$	$-2.053 \cdot 10^{-6}$	$-1.389 \cdot 10^{-5}$	$-5.434 \cdot 10^{-1}$

Example: Climate Index for Athens, Greece

RAD = 190
DD20 = 186

then:

CI   = 1.644

Step 2 Use of CI

To estimate roof pond performance for best case roof ponds with continuous spraying and daytime cover, enter the value of CI on the *x*-axis of any of the graphs on opposite page and read *y*-axis value for the roof type that more closely resembles that being considered. The graphs below illustrate this with the CI calculated above for Athens and the L roof type.

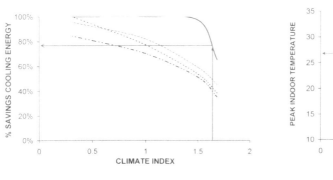
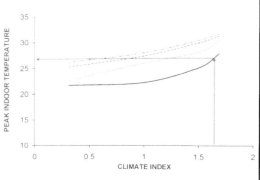

Similar data to those on the maps shown previously are represented on the adjacent graphs for Roof Ponds as a function of a Climate Index (CI). The CI of a location can be calculated using the empirical algorithm given here. The CI value can be used for a rough estimate of likely performance. This is illustrated in the adjacent box with an example.

6 ANNEX

ROOF PONDS

COOLING ENERGY SAVINGS FOR RESIDENTIAL ROOF POND VARIANTS AND OPERATIONAL SCHEDULES

ROOF TYPE: L
REF COOLING -3278
Water in bags
Metal deck

POND DEPTH (m)					WATER CHANGES PER HOUR				DROP RADIUS (mm)				JET HEIGHT (m)				SCHEDULES		
0.1	0.2	0.3	0.4	0.5	0	1	2	3	0.5	1	2	3	0.5	1	1.3	1.5	C-N	N-D	C-D
																	-77.2%	-65.4%	-88.7%
-50.2%	-68.9%	-77.2%	-81.7%	-84.4%															
					-22.4%	-77.2%	-85.0%	-87.7%											
									-85.4%	-77.2%	-61.1%	-49.9%							
													-68.5%	-75.1%	-77.2%	-78.3%			

ROOF TYPE: LI
REF COOLING -3278

Water exposed
Waterproofing
Thermal insulation
Metal deck

POND DEPTH (m)					WATER CHANGES PER HOUR				DROP RADIUS (mm)				JET HEIGHT (m)				SCHEDULES		
0.1	0.2	0.3	0.4	0.5	0	1	2	3	0.5	1	2	3	0.5	1	1.3	1.5	C-N	N-D	C-D
																	-49.0%	-45.0%	-53.0%
-42.8%	-46.9%	-49.0%	-50.3%	-51.2%															
					-35.4%	-49.0%	-51.5%	-52.8%											
									-51.8%	-49.0%	-44.1%	-41.4%							
													-46.5%	-48.5%	-49.0%	-49.5%			

ROOF TYPE: LI+
REF COOLING -2907

as LI and additional
thermal insulation

POND DEPTH (m)					WATER CHANGES PER HOUR				DROP RADIUS (mm)				JET HEIGHT (m)				SCHEDULES		
0.1	0.2	0.3	0.4	0.5	0	1	2	3	0.5	1	2	3	0.5	1	1.3	1.5	C-N	N-D	C-D
																	-38.1%	-35.0%	-41.0%
-33.3%	-36.6%	-38.1%	-39.2%	-40.0%															
					-27.6%	-38.1%	-40.4%	-41.2%											
									-40.4%	-38.1%	-34.6%	-32.2%							
													-36.1%	-37.4%	-38.1%	-38.3%			

ROOF TYPE: HI
REF COOLING -2805

Water exposed
Waterproofing
Insulation
Masonry roof structure

POND DEPTH (m)					WATER CHANGES PER HOUR				DROP RADIUS (mm)				JET HEIGHT (m)				SCHEDULES		
0.1	0.2	0.3	0.4	0.5	0	1	2	3	0.5	1	2	3	0.5	1	1.3	1.5	C-N	N-D	C-D
																	-41.6%	-38.0%	-44.0%
-36.9%	-39.8%	-41.6%	-42.6%	-43.3%															
					-29.3%	-41.6%	-43.9%	-44.7%											
									-43.8%	-41.6%	-37.7%	-35.3%							
													-39.3%	-40.9%	-41.6%	-42.1%			

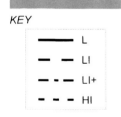

RESIDENTIAL COOLING ENERGY ATHENS

KEY
—— L
– – LI
–·– LI+
– ·· HI

The **reference** roof specifications and the predicted reference cooling energy demands are shown in the margin above for the four roof types introduced previously (L, LI, LI+ and HI; see key for graphs).

Results are given in the tables above for the energy that needs to be dissipated (indicated by negative sign) to maintain a constant 25°C for each value of the system design variables and operational conditions. The grey areas identify base case values of parameters.

% COOLING ENERGY SAVINGS FOR ROOF POND VARIANTS

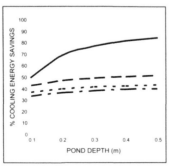
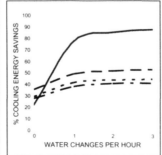
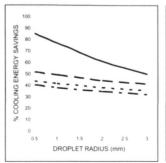
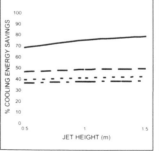
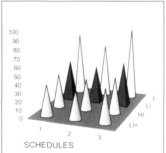

PEAK INDOOR TEMPERATURES & HOURS ABOVE 25°C FOR ROOF POND VARIANTS AND OPERATIONAL SCHEDULES

ROOF PONDS

ROOF TYPE: L

REF PEAK: 36.9°C
Water in bags
Metal deck

POND DEPTH (m)					WATER CHANGES PER HOUR				DROP RADIUS (mm)				JET HEIGHT (m)				SCHEDULES		
0.1	0.2	0.3	0.4	0.5	0.0	1.0	2.0	3.0	0.5	1.0	2.0	3.0	0.5	1.0	1.3	1.5	C-N	N-D	C-D
																	29.8	29.8	28.0
32.3	30.7	29.8	29.3	28.8															
					32.8	29.8	28.9	28.4											
									28.8	29.8	31.1	31.7							
													30.6	30.0	29.8	29.7			

ROOF TYPE: LI

REF PEAK: 36.9°C

Water exposed
Waterproofing
Thermal insulation
Metal deck

POND DEPTH (m)					WATER CHANGES PER HOUR				DROP RADIUS (mm)				JET HEIGHT (m)				SCHEDULES		
0.1	0.2	0.3	0.4	0.5	0.0	1.0	2.0	3.0	0.5	1.0	2.0	3.0	0.5	1.0	1.3	1.5	C-N	N-D	C-D
																	31.7	31.9	31.4
32.2	31.9	31.7	31.7	31.6															
					32.4	31.7	31.5	31.5											
									31.5	31.7	32.0	32.2							
													31.9	31.8	31.7	31.7			

ROOF TYPE: LI+

REF PEAK: 36.2°C

as **LI** and additional
thermal insulation

POND DEPTH (m)					WATER CHANGES PER HOUR				DROP RADIUS (mm)				JET HEIGHT (m)				SCHEDULES		
0.1	0.2	0.3	0.4	0.5	0.0	1.0	2.0	3.0	0.5	1.0	2.0	3.0	0.5	1.0	1.3	1.5	C-N	N-D	C-D
																	32.0	32.2	31.8
32.3	32.2	32.0	31.9	31.9															
					32.5	32.0	31.9	31.8											
									31.9	32.0	32.2	32.3							
													32.2	32.1	32.0	32.0			

ROOF TYPE: HI

REF PEAK: 33.8°C

Water exposed
Waterproofing
Insulation
Masonry roof

POND DEPTH (m)					WATER CHANGES PER HOUR				DROP RADIUS (mm)				JET HEIGHT (m)				SCHEDULES		
0.1	0.2	0.3	0.4	0.5	0.0	1.0	2.0	3.0	0.5	1.0	2.0	3.0	0.5	1.0	1.3	1.5	C-N	N-D	C-D
																	30.8	30.9	30.6
30.9	30.8	30.8	30.7	30.7															
					31.2	30.8	30.6	30.6											
									30.7	30.8	30.9	31.0							
													30.8	30.8	30.8	30.8			

The graphs opposite and below chart the % energy savings and the number of hours above 25°C for the four roof types and for the same roof pond parameters as the tables.

The **reference** roof specifications and the predicted reference peak indoor temperatures are shown in the right margin above for the four roof types introduced previously (L, LI, LI+ and HI; see key for graphs).

RESIDENTIAL PEAK INDOOR TEMPERATURE ATHENS

NUMBER OF HOURS INDOOR TEMPERATURE >25°C

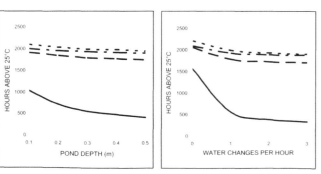

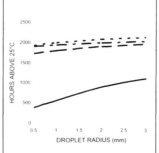

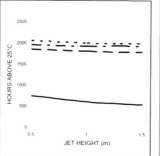

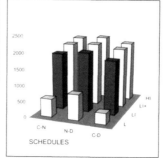

KEY

— L
– – LI
– · – LI+
- - - HI

ROOF PONDS

COOLING ENERGY DEMAND (kWh) AND % ENERGY SAVINGS FOR ROOF POND VARIANTS

ROOF TYPE: L
REF COOLING -947
Water in bags
Metal deck

POND DEPTH (m)					WATER CHANGES PER HOUR				DROP RADIUS (mm)				JET HEIGHT (m)				SCHEDULE-COVER SCHEDULE		
0.1	0.2	0.3	0.4	0.5	0.0	1.0	2.0	3.0	0.5	1.0	2.0	3.0	0.5	1.0	1.3	1.5	C-N	N-D	C-D
																	-97.6%	-100.0%	-100.0%
-73.2%	-92.6%	-97.6%	-99.1%	-99.6%															
					-56.5%	-97.6%	-99.5%	-99.9%											
									-99.6%	-97.6%	-89.4%	-81.7%							
													-93.8%	-96.8%	-97.6%	-98.0%			

ROOF TYPE: LI
REF COOLING -947
Water exposed
Waterproofing
Thermal insulation
Metal deck

POND DEPTH (m)					WATER CHANGES PER HOUR				DROP RADIUS (mm)				JET HEIGHT (m)				SCHEDULE-COVER SCHEDULE		
0.1	0.2	0.3	0.4	0.5	0.0	1.0	2.0	3.0	0.5	1.0	2.0	3.0	0.5	1.0	1.3	1.5	C-N	N-D	C-D
																	-76.2%	-82.0%	-82.6%
-68.5%	-73.5%	-76.2%	-77.5%	-78.6%															
					-61.9%	-76.2%	-78.7%	-79.6%											
									-78.8%	-76.2%	-71.6%	-69.0%							
													-73.4%	-75.4%	-76.2%	-76.3%			

ROOF TYPE: LI+
REF COOLING -757
as **LI** and additional
thermal insulation

POND DEPTH (m)					WATER CHANGES PER HOUR				DROP RADIUS (mm)				JET HEIGHT (m)				SCHEDULE-COVER SCHEDULE		
0.1	0.2	0.3	0.4	0.5	0.0	1.0	2.0	3.0	0.5	1.0	2.0	3.0	0.5	1.0	1.3	1.5	C-N	N-D	C-D
																	-64.6%	-70.4%	-70.9%
-57.4%	-62.2%	-64.6%	-65.8%	-66.9%															
					-51.1%	-64.6%	-67.1%	-68.0%											
									-67.1%	-64.6%	-60.4%	-57.4%							
													-62.1%	-64.0%	-64.6%	-65.0%			

ROOF TYPE: HI
REF COOLING -436
Water exposed
Waterproofing
Insulation
Masonry roof structure

POND DEPTH (m)					WATER CHANGES PER HOUR				DROP RADIUS (mm)				JET HEIGHT (m)				SCHEDULE-COVER SCHEDULE		
0.1	0.2	0.3	0.4	0.5	0.0	1.0	2.0	3.0	0.5	1.0	2.0	3.0	0.5	1.0	1.3	1.5	C-N	N-D	C-D
																	-71.4%	-77.3%	-77.5%
-65.0%	-69.0%	-71.4%	-72.5%	-74.5%															
					-51.7%	-71.4%	-74.6%	-75.8%											
									-74.8%	-71.4%	-65.8%	-61.9%							
													-68.4%	-70.8%	-71.4%	-72.2%			

RESIDENTIAL COOLING ENERGY MADRID

The **reference** roof specifications and the predicted reference cooling energy demands are shown in the margin above for the four roof types introduced previously (L, LI, LI+ and HI; see key for graphs).

Results are given in the tables above for the energy that needs to be dissipated (indicated by negative sign) to maintain a constant 25°C for each value of the system design variables and operational conditions. The grey areas identify base case values of parameters.

KEY

——— L
— — LI
—·— LI+
- - - HI

% COOLING ENERGY SAVINGS FOR ROOF POND VARIANTS

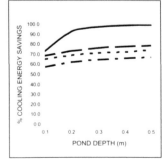
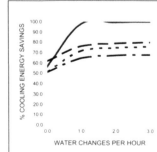
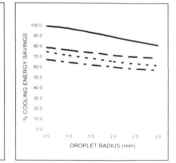
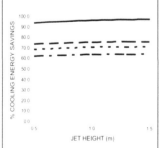
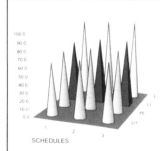

PEAK INDOOR TEMPERATURES & NUMBER OF HOURS ABOVE 25°C FOR ROOF POND VARIANTS

ROOF PONDS

ROOF TYPE: L

POND DEPTH (m)					WATER CHANGES PER HOUR				DROP RADIUS (mm)				JET HEIGHT (m)				SCHEDULES		
0.1	0.2	0.3	0.4	0.5	0.0	1.0	2.0	3.0	0.5	1.0	2.0	3.0	0.5	1.0	1.3	1.5	C-N	N-D	C-D
																	25.0	22.8	22.3
27.9	26.1	25.0	24.4	24.0															
					28.9	25.0	24.1	23.7											
									24.0	25.0	26.5	27.3							
													25.9	25.3	25.0	24.9			

REF PEAK: 34.3°C
Water in bags
Metal deck

ROOF TYPE: LI

POND DEPTH (m)					WATER CHANGES PER HOUR				DROP RADIUS (mm)				JET HEIGHT (m)				SCHEDULES		
0.1	0.2	0.3	0.4	0.5	0.0	1.0	2.0	3.0	0.5	1.0	2.0	3.0	0.5	1.0	1.3	1.5	C-N	N-D	C-D
																	28.1	27.6	27.5
28.7	28.3	28.1	28.0	27.9															
					29.1	28.1	27.9	27.8											
									27.9	28.1	28.4	28.7							
													28.3	28.1	28.1	28.1			

REF PEAK: 34.3°C

Water exposed
Waterproofing
Thermal insulation
Metal deck

ROOF TYPE: LI+

POND DEPTH (m)					WATER CHANGES PER HOUR				DROP RADIUS (mm)				JET HEIGHT (m)				SCHEDULES		
0.1	0.2	0.3	0.4	0.5	0.0	1.0	2.0	3.0	0.5	1.0	2.0	3.0	0.5	1.0	1.3	1.5	C-N	N-D	C-D
																	28.5	28.1	28.1
29.0	28.6	28.5	28.4	28.3															
					29.3	28.5	28.3	28.3											
									28.3	28.5	28.8	28.9							
													28.6	28.6	28.5	28.5			

REF PEAK: 33.6°C

as **LI** and additional
thermal insulation

ROOF TYPE: HI

POND DEPTH (m)					WATER CHANGES PER HOUR				DROP RADIUS (mm)				JET HEIGHT (m)				SCHEDULES		
0.1	0.2	0.3	0.4	0.5	0.0	1.0	2.0	3.0	0.5	1.0	2.0	3.0	0.5	1.0	1.3	1.5	C-N	N-D	C-D
																	26.3	26.1	26.1
26.6	26.4	26.3	26.3	26.2															
					27.0	26.3	26.2	26.2											
									26.2	26.3	26.6	26.7							
													26.5	26.4	26.3	26.2			

REF PEAK: 30.1°C

Water exposed
Waterproofing
Insulation
Masonry roof structure

The graphs opposite and below chart the % energy savings and the number of hours above 25°C for the four roof types and for the same roof pond parameters as the tables.

The **reference** roof specifications and the predicted reference peak indoor temperatures are shown in the right margin above for the four roof types introduced previously (L, LI, LI+ and HI; see key for graphs).

RESIDENTIAL PEAK INDOOR TEMPERATURE MADRID

NUMBER OF HOURS INDOOR TEMPERATURE >25°C

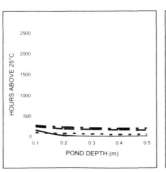

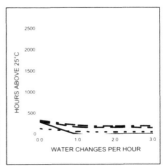

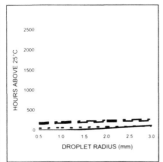

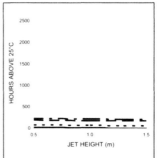

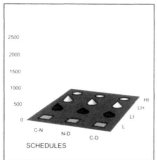

KEY

— L
– – LI
–·– LI+
- - - HI

ROOF POND & COOLING PANEL

CONTRIBUTION FROM ROOF PONDS WITH COOLING PANELS ON RESIDENTIAL BUILDINGS

% COOLING ENERGY SAVING   **Seville**

ROOF TYPE: **L**
REF COOLING: **-5590**

DISTANCE BETWEEN PIPES (m)			WATER FLOW (kg/s)			SCHEDULE		
0.1	0.2	0.3	6	12	18	Continuous	Night	Daytime
-65.6%	-68.1%	-64.4%						
			-67.6%	-68.1%	-68.7%			
						-68.1%	-77.1%	-75.8%

ROOF TYPE: **LI**
REF COOLING: **-5590**

DISTANCE BETWEEN PIPES (m)			WATER FLOW (kg/s)			SCHEDULE		
0.1	0.2	0.3	6	12	18	Continuous	Night	Daytime
-36.5%	-33.9%	-24.8%						
			-33.8%	-33.9%	-34.1%			
						-33.9%	-29.6%	-35.0%

ROOF TYPE: **LI+**
REF COOLING: **-5326**

DISTANCE BETWEEN PIPES (m)			WATER FLOW (kg/s)			SCHEDULE		
0.1	0.2	0.3	6	12	18	Continuous	Night	Daytime
-31.9%	-28.8%	-19.0%						
			-28.7%	-28.8%	-29.0%			
						-28.8%	-22.8%	-28.9%

ROOF TYPE: **HI**
REF COOLING: **-5326**

DISTANCE BETWEEN PIPES (m)			WATER FLOW (kg/s)			SCHEDULE		
0.1	0.2	0.3	6	12	18	Continuous	Night	Daytime
-35.6%	-32.8%	-23.0%						
			-32.6%	-32.8%	-33.0%			
						-32.8%	-27.8%	-32.3%

SEVILLE

NUMBER OF HOURS INDOOR TEMPERATURE >25°C   **Seville**

ROOF TYPE: **L**
REF NO. HOURS: **1750**

DISTANCE BETWEEN PIPES (m)			WATER FLOW (kg/s)			SCHEDULE		
0.1	0.2	0.3	6	12	18	Continuous	Night	Daytime
701	635	717						
			649	635	618			
						635	421	460

ROOF TYPE: **LI**
REF NO. HOURS: **1750**

DISTANCE BETWEEN PIPES (m)			WATER FLOW (kg/s)			SCHEDULE		
0.1	0.2	0.3	6	12	18	Continuous	Night	Daytime
1506	1564	1728						
			1557	1564	1567			
						1564	1565	1569

ROOF TYPE: **LI+**
REF NO. HOURS: **1751**

DISTANCE BETWEEN PIPES (m)			WATER FLOW (kg/s)			SCHEDULE		
0.1	0.2	0.3	6	12	18	Continuous	Night	Daytime
1540	1600	1764						
			1594	1600	1604			
						1600	1615	1639

ROOF TYPE: **HI**
REF NO. HOURS: **1819**

DISTANCE BETWEEN PIPES (m)			WATER FLOW (kg/s)			SCHEDULE		
0.1	0.2	0.3	6	12	18	Continuous	Night	Daytime
1462	1533	1730						
			1522	1533	1536			
						1533	1598	1564

On multi-storey buildings a panel that is thermally coupled with the ceiling of occupied spaces can act as an interim heat sink and as a heat exchanger. The use of a small pump to circulate water to and from a roof pond has a negligible economic and energy cost.

The total cooling potential available is limited by the size of the roof pond. The tables show results for a two-storey residential building with the same size roof pond as on previous tables. The total cooling load is about twice as high, thus the percentage savings are lower than for the roof pond alone on a single-storey building. It can be seen, however, that in the best case combinations the savings exceed 60% of the cooling demand.

CONTRIBUTION FROM COOLING PANELS (WITHOUT ROOF POND) ON RESIDENTIAL BUILDINGS

Given a source of fresh water supply (for example, a nearby river or lake) cooling panels can contribute to the cooling of buildings without fitting a roof pond. The tables show results for cooling panels of different water flow and pipe design characteristics supplied from sources with water temperatures in the range 18–22°C and operating continuously, day-time only or night-time only. With continuous operation the energy savings compared with air-conditioning rise to 40–50%.

% COOLING ENERGY SAVINGS FROM COOLING PANELS WITH FRESH WATER SUPPLY — Seville

ROOF TYPE: LI   REF COOLING: -5590

DISTANCE BETWEEN PIPES (m)			WATER FLOW (kg/s)			WATER TEMPERATURE (°C)			SCHEDULE		
0.1	0.2	0.3	6	12	18	18	20	22	Continuous	Night	Daytime
-49.5%	-38.5%	-24.3%									
			-38.4%	-38.5%	-38.6%						
						-51.9%	-38.5%	-23.3%			
									-38.5%	-8.0%	-24.3%

ROOF TYPE: LI+   REF COOLING: -5326

DISTANCE BETWEEN PIPES (m)			WATER FLOW (kg/s)			WATER TEMPERATURE (°C)			SCHEDULE		
0.1	0.2	0.3	6	12	18	18	20	22	Continuous	Night	Daytime
-52.8%	-41.7%	-27.3%									
			-41.6%	-41.7%	-41.8%						
						-55.8%	-41.7%	-25.4%			
									-41.7%	-10.3%	-26.3%

ROOF TYPE: HI   REF COOLING: -5241

DISTANCE BETWEEN PIPES (m)			WATER FLOW (kg/s)			WATER TEMPERATURE (°C)			SCHEDULE		
0.1	0.2	0.3	6	12	18	18	20	22	Continuous	Night	Daytime
-63.2%	-51.6%	-36.1%									
			51.4%	-51.6%	-51.7%						
						-66.5%	-51.6%	-33.6%			
									-51.6%	-19.5%	-28.8%

NUMBER OF HOURS INDOOR TEMPERATURE >25°C — Seville

ROOF TYPE: LI   REF NO. HOURS: 1750

DISTANCE BETWEEN PIPES (m)			WATER FLOW (kg/s)			WATER TEMPERATURE (°C)			SCHEDULE		
0.1	0.2	0.3	6	12	18	18	20	22	Continuous	Night	Daytime
1043	1271	1539									
			1274	1271	1271						
						1016	1271	1526			
									1271	1601	1685

ROOF TYPE: LI+   REF NO. HOURS: 1751

DISTANCE BETWEEN PIPES (m)			WATER FLOW (kg/s)			WATER TEMPERATURE (°C)			SCHEDULE		
0.1	0.2	0.3	6	12	18	18	20	22	Continuous	Night	Daytime
985	1216	1498									
			1219	1216	1216						
						940	1216	1495			
									1216	1573	1643

ROOF TYPE: HI   REF NO. HOURS: 1819

DISTANCE BETWEEN PIPES (m)			WATER FLOW (kg/s)			WATER TEMPERATURE (°C)			SCHEDULE		
0.1	0.2	0.3	6	12	18	18	20	22	Continuous	Night	Daytime
787	1059	1439									
			1062	1059	1056						
						735	1059	1444			
									1059	1656	1610

ROOF POND & COOLING PANEL

CONTRIBUTION FROM ROOF PONDS WITH COOLING PANELS ON RESIDENTIAL BUILDINGS

% COOLING ENERGY SAVING **Athens**

ROOF TYPE: **L**
REF COOLING: **-7513**

DISTANCE BETWEEN PIPES (m)			WATER FLOW (kg/s)			SCHEDULE		
0.1	0.2	0.3	6	12	18	Continuous	Night	Daytime
-53.0%	-56.0%	-51.5%						
			-55.5%	-56.0%	-56.5%			
						-56.0%	-67.4%	-64.0%

ROOF TYPE: **LI**
REF COOLING: **-7513**

DISTANCE BETWEEN PIPES (m)			WATER FLOW (kg/s)			SCHEDULE		
0.1	0.2	0.3	6	12	18	Continuous	Night	Daytime
-29.5%	-27.7%	-20.0%						
			-27.4%	-27.7%	-27.9%			
						-27.7%	-25.2%	-27.5%

ROOF TYPE: **LI+**
REF COOLING: **-7171**

DISTANCE BETWEEN PIPES (m)			WATER FLOW (kg/s)			SCHEDULE		
0.1	0.2	0.3	6	12	18	Continuous	Night	Daytime
-25.0%	-22.9%	-14.7%						
			-22.7%	-22.9%	-23.2%			
						-22.9%	-19.1%	-22.0%

ROOF TYPE: **HI**
REF COOLING: **-7171**

DISTANCE BETWEEN PIPES (m)			WATER FLOW (kg/s)			SCHEDULE		
0.1	0.2	0.3	6	12	18	Continuous	Night	Daytime
-30.9%	-29.1%	-21.2%						
			-28.9%	-29.1%	-29.3%			
						-29.1%	-26.1%	-27.7%

ATHENS

NUMBER OF HOURS INDOOR TEMPERATURE >25°C **Athens**

ROOF TYPE: **L**
REF NO. HOURS: **2534**

DISTANCE BETWEEN PIPES (m)			WATER FLOW (kg/s)			SCHEDULE		
0.1	0.2	0.3	6	12	18	Continuous	Night	Daytime
1838	1715	1811						
			1659	1715	1744			
						1715	1192	1351

ROOF TYPE: **LI**
REF NO. HOURS: **2534**

DISTANCE BETWEEN PIPES (m)			WATER FLOW (kg/s)			SCHEDULE		
0.1	0.2	0.3	6	12	18	Continuous	Night	Daytime
2438	2471	2574						
			2466	2471	2471			
						2471	2468	2500

ROOF TYPE: **LI+**
REF NO. HOURS: **2537**

DISTANCE BETWEEN PIPES (m)			WATER FLOW (kg/s)			SCHEDULE		
0.1	0.2	0.3	6	12	18	Continuous	Night	Daytime
2456	2490	2590						
			2486	2490	2491			
						2490	2500	2529

ROOF TYPE: **HI**
REF NO. HOURS: **2563**

DISTANCE BETWEEN PIPES (m)			WATER FLOW (kg/s)			SCHEDULE		
0.1	0.2	0.3	6	12	18	Continuous	Night	Daytime
2426	2461	2570						
			2462	2461	2457			
						2461	2503	2490

On multi-storey buildings a panel that is thermally coupled with the ceiling of occupied spaces can act as an interim heat sink and as a heat exchanger. The use of a small pump to circulate water to and from a roof pond has a negligible economic and energy cost.

The total cooling potential available is limited by the size of the roof pond. The tables show results for a two-storey residential building with the same size roof pond as on previous tables. The total cooling load is about twice as high, thus the percentage savings are lower than for the roof pond alone on a single-storey building. It can be seen, however, that in the best case combinations the savings exceed 60% of the cooling demand.

CONTRIBUTION FROM COOLING PANELS (WITHOUT ROOF POND) ON RESIDENTIAL BUILDINGS

Given a source of fresh water supply (for example, a nearby river or lake) cooling panels can contribute to the cooling of buildings without fitting a roof pond. The tables show results for cooling panels of different water flow and pipe design characteristics supplied from sources with water temperatures in the range 18–22°C and operating continuously, daytime only or night-time only. With continuous operation the energy savings compared with air-conditioning rise to 40–50%.

% COOLING ENERGY SAVINGS FROM COOLING PANELS WITH FRESH WATER SUPPLY — Athens

ROOF TYPE: **LI** — REF COOLING: **-7513**

DISTANCE BETWEEN PIPES (m)			WATER FLOW (kg/s)			WATER TEMPERATURE (°C)			SCHEDULE		
0.1	0.2	0.3	6	12	18	18	20	22	Continuous	Night	Daytime
-53.0%	-41.5%	-26.9%									
			-41.3%	-41.5%	-41.5%						
						-54.7%	-41.5%	-26.5%			
									-41.5%	-9.8%	-22.1%

ROOF TYPE: **LI+** — REF COOLING: **-7171**

DISTANCE BETWEEN PIPES (m)			WATER FLOW (kg/s)			WATER TEMPERATURE (°C)			SCHEDULE		
0.1	0.2	0.3	6	12	18	18	20	22	Continuous	Night	Daytime
-56.0%	-44.3%	-29.6%									
			-44.1%	-44.3%	-44.3%						
						-58.2%	-44.3%	-28.3%			
									-44.3%	-12.3%	-23.7%

ROOF TYPE: **HI** — REF COOLING: **-7571**

DISTANCE BETWEEN PIPES (m)			WATER FLOW (kg/s)			WATER TEMPERATURE (°C)			SCHEDULE		
0.1	0.2	0.3	6	12	18	18	20	22	Continuous	Night	Daytime
-61.5%	-50.1%	-35.8%									
			-49.9%	-50.1%	-50.1%						
						-63.9%	-50.1%	-34.4%			
									-50.1%	-21.9%	-26.4%

NUMBER OF HOURS INDOOR TEMPERATURE >25ºC — Athens

ROOF TYPE: **LI** — REF NO. HOURS: **2534**

DISTANCE BETWEEN PIPES (m)			WATER FLOW (kg/s)			WATER TEMPERATURE (°C)			SCHEDULE		
0.1	0.2	0.3	6	12	18	18	20	22	Continuous	Night	Daytime
1708	2044	2328									
			2048	2044	2042						
						1706	2044	2301			
									2044	2406	2452

ROOF TYPE: **LI+** — REF NO. HOURS: **2537**

DISTANCE BETWEEN PIPES (m)			WATER FLOW (kg/s)			WATER TEMPERATURE (°C)			SCHEDULE		
0.1	0.2	0.3	6	12	18	18	20	22	Continuous	Night	Daytime
1651	2004	2308									
			2005	2004	2001						
						1632	2004	2288			
									2004	2395	2441

ROOF TYPE: **HI** — REF NO. HOURS: **2563**

DISTANCE BETWEEN PIPES (m)			WATER FLOW (kg/s)			WATER TEMPERATURE (°C)			SCHEDULE		
0.1	0.2	0.3	6	12	18	18	20	22	Continuous	Night	Daytime
1607	1997	2348									
			2000	1997	1994						
						1586	1997	2333			
									1997	2472	2470

ROOF POND & COOLING PANEL

CONTRIBUTION FROM ROOF PONDS WITH COOLING PANELS ON RESIDENTIAL BUILDINGS

% COOLING ENERGY SAVING   **Madrid**

ROOF TYPE: L
REF COOLING: -3161

DISTANCE BETWEEN PIPES (m)			WATER FLOW (kg/s)			SCHEDULE		
0.1	0.2	0.3	6.0	12.0	18.0	Continuous	Night	Daytime
-89.8%	-91.2%	-89.8%						
			-90.1%	-91.2%	-91.8%			
						-91.2%	-95.6%	-94.9%

ROOF TYPE: LI
REF COOLING: -3161

DISTANCE BETWEEN PIPES (m)			WATER FLOW (kg/s)			SCHEDULE		
0.1	0.2	0.3	6.0	12.0	18.0	Continuous	Night	Daytime
-56.0%	-52.1%	-40.8%						
			-51.6%	-52.1%	-52.5%			
						-52.1%	-45.3%	-53.5%

ROOF TYPE: LI+
REF COOLING: -2976

DISTANCE BETWEEN PIPES (m)			WATER FLOW (kg/s)			SCHEDULE		
0.1	0.2	0.3	6.0	12.0	18.0	Continuous	Night	Daytime
-51.1%	-46.3%	-33.6%						
			-45.7%	-46.3%	-46.8%			
						-46.3%	-36.3%	-46.4%

ROOF TYPE: HI
REF COOLING: -2976

DISTANCE BETWEEN PIPES (m)			WATER FLOW (kg/s)			SCHEDULE		
0.1	0.2	0.3	6.0	12.0	18.0	Continuous	Night	Daytime
-53.8%	-49.2%	-35.3%						
			-48.7%	-49.2%	-49.6%			
						-49.2%	-40.8%	-48.4%

MADRID

NUMBER OF HOURS INDOOR TEMPERATURE >25ºC   **Madrid**

ROOF TYPE: L
REF NO. HOURS: 1219

DISTANCE BETWEEN PIPES (m)			WATER FLOW (kg/s)			SCHEDULE		
0.1	0.2	0.3	6.0	12.0	18.0	Continuous	Night	Daytime
94	72	88						
			101	72	63			
						72	41	44

ROOF TYPE: LI
REF NO. HOURS: 1219

DISTANCE BETWEEN PIPES (m)			WATER FLOW (kg/s)			SCHEDULE		
0.1	0.2	0.3	6.0	12.0	18.0	Continuous	Night	Daytime
751	817	1000						
			816	817	820			
						817	888	782

ROOF TYPE: LI+
REF NO. HOURS: 1216

DISTANCE BETWEEN PIPES (m)			WATER FLOW (kg/s)			SCHEDULE		
0.1	0.2	0.3	6.0	12.0	18.0	Continuous	Night	Daytime
790	866	1063						
			860	866	867			
						866	958	853

ROOF TYPE: HI
REF NO. HOURS: 1185

DISTANCE BETWEEN PIPES (m)			WATER FLOW (kg/s)			SCHEDULE		
0.1	0.2	0.3	6.0	12.0	18.0	Continuous	Night	Daytime
660	731	927						
			729	731	733			
						731	822	734

On multi-storey buildings a panel that is thermally coupled with the ceiling of occupied spaces can act as an interim heat sink and as a heat exchanger. The use of a small pump to circulate water to and from a roof pond has a negligible economic and energy cost.

The total cooling potential available is limited by the size of the roof pond. The tables show results for a two-storey residential building with the same size roof pond as on previous tables. The total cooling load is about twice as high, thus the percentage savings are lower than for the roof pond alone on a single-storey building. It can be seen, however, that in the best case combinations the savings exceed 60% of the cooling demand.

CONTRIBUTION FROM COOLING PANELS (WITHOUT ROOF POND) ON RESIDENTIAL BUILDINGS

Given a source of fresh water supply (for example, a nearby river or lake) cooling panels can contribute to the cooling of buildings without fitting a roof pond. The tables show results for cooling panels of different water flow and pipe design characteristics supplied from sources with water temperatures in the range 18–22°C and operating continuously, daytime only or night-time only. With continuous operation the energy savings compared with air-conditioning rise to 40–50%.

% COOLING ENERGY SAVINGS FROM COOLING PANELS WITH FRESH WATER SUPPLY — Madrid

ROOF TYPE: **LI** — REF COOLING: **-3161**

DISTANCE BETWEEN PIPES (m)			WATER FLOW (kg/s)			WATER TEMPERATURE (°C)			SCHEDULE		
0.1	0.2	0.3	6	12	18	18	20	22	Continuous	Night	Daytime
-54.3%	-43.1%	-27.5%									
			-42.9%	-43.1%	-43.1%						
						-57.5%	-43.1%	-25.2%			
									-43.1%	-4.5%	-30.7%

ROOF TYPE: **LI+** — REF COOLING: **-2976**

DISTANCE BETWEEN PIPES (m)			WATER FLOW (kg/s)			WATER TEMPERATURE (°C)			SCHEDULE		
0.1	0.2	0.3	6	12	18	18	20	22	Continuous	Night	Daytime
-58.4%	-46.9%	-31.0%									
			-46.8%	-46.9%	-47.0%						
						-62.1%	-46.9%	-27.7%			
									-46.9%	-7.3%	32.7%

ROOF TYPE: **HI** — REF COOLING: **-3161**

DISTANCE BETWEEN PIPES (m)			WATER FLOW (kg/s)			WATER TEMPERATURE (°C)			SCHEDULE		
0.1	0.2	0.3	6	12	18	18	20	22	Continuous	Night	Daytime
-54.3%	-43.1%	-27.5%									
			-42.9%	-43.1%	-43.1%						
						57.5%	-43.1%	-25.2%			
									-43.1%	-4.5%	-30.7%

NUMBER OF HOURS INDOOR TEMPERATURE >25°C — Madrid

ROOF TYPE: **LI** — REF NO. HOURS: **1219**

DISTANCE BETWEEN PIPES (m)			WATER FLOW (kg/s)			WATER TEMPERATURE (°C)			SCHEDULE		
0.1	0.2	0.3	6	12	18	18	20	22	Continuous	Night	Daytime
628	812	1031									
			813	812	811						
						606	812	1033			
									812	1222	993

ROOF TYPE: **LI+** — REF NO. HOURS: **1216**

DISTANCE BETWEEN PIPES (m)			WATER FLOW (kg/s)			WATER TEMPERATURE (°C)			SCHEDULE		
0.1	0.2	0.3	6	12	18	18	20	22	Continuous	Night	Daytime
576	761	980									
			762	761	761						
						536	761	998			
									761	1181	962

ROOF TYPE: **HI** — REF NO. HOURS: **1219**

DISTANCE BETWEEN PIPES (m)			WATER FLOW (kg/s)			WATER TEMPERATURE (°C)			SCHEDULE		
0.1	0.2	0.3	6	12	18	18	20	22	Continuous	Night	Daytime
628	812	1031									
			813	812	811						
						606	812	1033			
									812	1222	993

AIR RADIATORS

COOLING ENERGY SAVINGS % WITH AIR RADIATORS

COOLING ENERGY DEMAND, kWh — Seville

ROOF TYPE: **LI**
REF COOLING: **-2394**

LENGTH (m)					CHANNEL THICKNESS (m)			AIR FLOW (kg/m²)					SCHEDULE		
2	4	6	8	10	0.01	0.03	0.05	0.03	0.05	0.07	0.08	0.1	N	D	C
-2.5%	-3.6%	-4.3%	-5.0%	-5.3%											
					-5.5%	-4.3%	-3.7%								
								-4.3%	-4.9%	-4.9%	-4.7%	-4.2%			
													-4.3%	34.9%	30.2%

ROOF TYPE: **LI+**
REF COOLING: **-2084**

LENGTH (m)					CHANNEL THICKNESS (m)			AIR FLOW (kg/m²)					SCHEDULE		
2	4	6	8	10	0.01	0.03	0.05	0.03	0.05	0.07	0.08	0.1	N	D	C
-3.3%	-4.3%	-5.3%	-5.8%	-6.5%											
					-6.4%	-5.3%	-4.6%								
								-5.3%	-6.0%	-6.2%	-6.1%	-5.2%			
													-5.3%	-40.5%	-35.3%

ROOF TYPE: **HI**
REF COOLING: **-1718**

LENGTH (m)					CHANNEL THICKNESS (m)			AIR FLOW (kg/m²)					SCHEDULE		
2	4	6	8	10	0.01	0.03	0.05	0.03	0.05	0.07	0.08	0.1	N	D	C
-9.4%	-11.4%	-13.1%	-13.6%	-14.5%											
					-14.4%	-13.1%	-11.6%								
								-13.1%	-16.0%	-17.6%	-17.9%	-18.3%			
													-13.1%	48.7%	33.8%

SEVILLE

The results for Seville are more promising with energy savings of 5–15% as a function of optimising length of radiator and schedule of operation. There is also a possible reduction of 15–30% in the hours above 25°C and a drop by up to 2K in peak indoor temperature.

% COOLING ENERGY SAVINGS FROM AIR RADIATORS

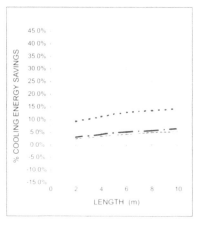

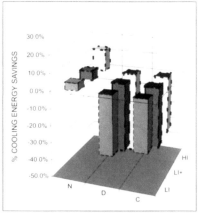

THERMAL COMFORT IMPROVEMENTS WITH AIR RADIATORS

AIR RADIATORS

NO. OF HOURS INDOORS ABOVE 25°C WITHOUT AC — Seville

ROOF TYPE: LI — REF COOLING: 1723

LENGTH (m)					CHANNEL THICKNESS (m)			AIR FLOW (kg/m²)					SCHEDULE		
2	4	6	8	10	0.01	0.03	0.05	0.03	0.05	0.07	0.08	0.1	N	D	C
1525	1500	1481	1465	1456											
					1457	1481	1499								
								1481	1331	1233	1199	1145			
													1481	2098	1612

ROOF TYPE: LI+ — REF COOLING: 1706

LENGTH (m)					CHANNEL THICKNESS (m)			AIR FLOW (kg/m²)					SCHEDULE		
2	4	6	8	10	0.01	0.03	0.05	0.03	0.05	0.07	0.08	0.1	N	D	C
1414	1389	1374	1353	1337											
					1340	1374	1381								
								1374	1212	1109	1072	1019			
													1374	2106	1566

ROOF TYPE: HI — REF COOLING: 1765

LENGTH (m)					CHANNEL THICKNESS (m)			AIR FLOW (kg/m²)					SCHEDULE		
2	4	6	8	10	0.01	0.03	0.05	0.03	0.05	0.07	0.08	0.1	N	D	C
1353	1322	1298	1271	1252											
					1252	1298	1311								
								1298	1086	949	907	827			
													1298	2360	1573

SEVILLE

PEAK INDOOR TEMPERATURE WITHOUT AC, °C — Seville

ROOF TYPE: LI — REF NO. HOURS: 37

LENGTH (m)					CHANNEL THICKNESS (m)			AIR FLOW (kg/m²)					SCHEDULE		
2	4	6	8	10	0.01	0.03	0.05	0.03	0.05	0.07	0.08	0.1	N	D	C
34.4	34.3	34.3	34.3	34.3											
					34.3	34.3	34.3								
								34.3	34.1	34.0	33.9	33.9			
													34.3	36.2	35.7

ROOF TYPE: LI+ — REF NO. HOURS: 36

LENGTH (m)					CHANNEL THICKNESS (m)			AIR FLOW (kg/m²)					SCHEDULE		
2	4	6	8	10	0.01	0.03	0.05	0.03	0.05	0.07	0.08	0.1	N	D	C
33.4	33.4	33.3	33.4	33.3											
					33.3	33.3	33.3								
								33.3	33.1	33.1	33.0	32.8			
													33.3	35.5	35.0

ROOF TYPE: HI — REF NO. HOURS: 34

LENGTH (m)					CHANNEL THICKNESS (m)			AIR FLOW (kg/m²)					SCHEDULE		
2	4	6	8	10	0.01	0.03	0.05	0.03	0.05	0.07	0.08	0.1	N	D	C
30.0	29.9	29.9	29.9	29.8											
					29.9	29.9	29.9								
								29.9	29.7	29.6	29.5	29.6			
													29.9	32.5	31.9

AIR RADIATORS

COOLING ENERGY DEMAND (kWh) WITH AIR RADIATORS

COOLING ENERGY DEMAND, kWh **Athens**

ROOF TYPE: **LI**
REF COOLING: **-3278**

LENGTH (m)					CHANNEL THICKNESS (m)			AIR FLOW (kg/m²)					SCHEDULE		
2	4	6	8	10	0.01	0.03	0.05	0.03	0.05	0.07	0.08	0.1	N	D	C
3.2%	2.7%	2.4%	2.0%	1.5%											
					1.6%	2.4%	2.7%								
								2.4%	4.9%	8.1%	9.8%	13.0%			
													2.4%	28.2%	33.4%

ROOF TYPE: **LI+**
REF COOLING: **-2907**

LENGTH (m)					CHANNEL THICKNESS (m)			AIR FLOW (kg/m²)					SCHEDULE		
2	4	6	8	10	0.01	0.03	0.05	0.03	0.05	0.07	0.08	0.1	N	D	C
3.7%	3.0%	2.4%	2.0%	1.6%											
					1.7%	2.4%	3.2%								
								2.4%	5.5%	9.0%	10.7%	14.5%			
													2.4%	31.4%	37.0%

ROOF TYPE: **HI**
REF COOLING: **-2805**

LENGTH (m)					CHANNEL THICKNESS (m)			AIR FLOW (kg/m²)					SCHEDULE		
2	4	6	8	10	0.01	0.03	0.05	0.03	0.05	0.07	0.08	0.1	N	D	C
1.9%	1.0%	0.4%	0.0%	-0.5%											
					-0.5%	0.4%	1.1%								
								0.4%	2.5%	5.6%	7.0%	10.5%			
													0.4%	32.4%	35.4%

ATHENS

The results suggest that care is required to ensure applicability. In the case of Athens the results are not promising. What the negative energy savings seem to indicate is that the air radiator cannot lower the building temperature to the air-conditioning setting of 25°C at any time of day or night in this location. There are marginal reductions in the number of hours above 25°C and a lowering by about 2K of the peak indoor temperatures in the best cases.

% COOLING ENERGY SAVINGS FROM AIR RADIATORS

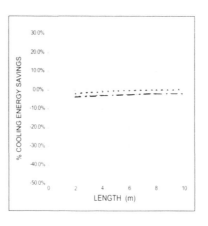

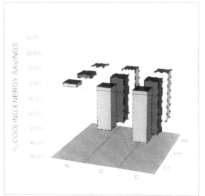

THERMAL COMFORT IMPROVEMENTS WITH AIR RADIATORS

AIR RADIATORS

NO. OF HOURS INDOORS ABOVE 25°C WITHOUT AC — Athens

ROOF TYPE: LI
REF COOLING: 2390

LENGTH (m)					CHANNEL THICKNESS (m)			AIR FLOW (kg/m²)					SCHEDULE		
2	4	6	8	10	0.01	0.03	0.05	0.03	0.05	0.07	0.08	0.1	N	D	C
2331	2318	2309	2296	2293											
					2293	2309	2315								
								2309	2217	2150	2123	2087			
													2309	2527	2312

ROOF TYPE: LI+
REF COOLING: 2390

LENGTH (m)					CHANNEL THICKNESS (m)			AIR FLOW (kg/m²)					SCHEDULE		
2	4	6	8	10	0.01	0.03	0.05	0.03	0.05	0.07	0.08	0.1	N	D	C
2283	2272	2259	2248	2243											
					2243	2259	2266								
								2259	2155	2081	2051	2015			
													2259	2517	2295

ROOF TYPE: HI
REF COOLING: 2506

LENGTH (m)					CHANNEL THICKNESS (m)			AIR FLOW (kg/m²)					SCHEDULE		
2	4	6	8	10	0.01	0.03	0.05	0.03	0.05	0.07	0.08	0.1	N	D	C
2446	2441	2436	2433	2434											
					2433	2436	2440								
								2436	2333	2242	2207	2148			
													2436	2567	2460

ATHENS

PEAK INDOOR TEMPERATURE WITHOUT AC, °C — Athens

ROOF TYPE: LI
REF NO. HOURS: 37

LENGTH (m)					CHANNEL THICKNESS (m)			AIR FLOW (kg/m²)					SCHEDULE		
2	4	6	8	10	0.01	0.03	0.05	0.03	0.05	0.07	0.08	0.1	N	D	C
34.9	34.8	38.8	34.8	34.7											
					34.8	34.8	34.8								
								34.8	34.7	34.7	34.6	34.6			
													34.8	35.5	35.3

ROOF TYPE: LI+
REF NO. HOURS: 36

LENGTH (m)					CHANNEL THICKNESS (m)			AIR FLOW (kg/m²)					SCHEDULE		
2	4	6	8	10	0.01	0.03	0.05	0.03	0.05	0.07	0.08	0.1	N	D	C
34.0	33.9	33.9	33.9	33.9											
					33.9	33.9	33.9								
								33.9	33.8	33.8	33.8	33.7			
													33.9	34.9	34.7

ROOF TYPE: HI
REF NO. HOURS: 34

LENGTH (m)					CHANNEL THICKNESS (m)			AIR FLOW (kg/m²)					SCHEDULE		
2	4	6	8	10	0.01	0.03	0.05	0.03	0.05	0.07	0.08	0.1	N	D	C
32.1	32.1	32.1	32.1	32.1											
					32.1	32.1	32.1								
								32.1	32.0	31.9	31.9	31.8			
													32.1	33.4	33.1

AIR RADIATORS

COOLING ENERGY DEMAND (kWh) WITH AIR RADIATORS

COOLING ENERGY DEMAND, kWh   **Madrid**

ROOF TYPE: **LI**
REF COOLING: **-947**

LENGTH (m)					CHANNEL THICKNESS (m)			AIR FLOW (kg/m²)					SCHEDULE		
2	4	6	8	10	0.01	0.03	0.05	0.03	0.05	0.07	0.08	0.1	N	D	C
-7.2%	-8.8%	-9.6%	-10.3%	-10.7%											
					-10.6%	-9.6%	-8.8%								
								-9.6%	-12.1%	-13.5%	-14.2%	-14.7%			
													-9.6%	32.1%	23.6%

ROOF TYPE: **LI+**
REF COOLING: **-757**

LENGTH (m)					CHANNEL THICKNESS (m)			AIR FLOW (kg/m²)					SCHEDULE		
2	4	6	8	10	0.01	0.03	0.05	0.03	0.05	0.07	0.08	0.1	N	D	C
-8.6%	-10.4%	-11.3%	-11.8%	-12.7%											
					-12.7%	-11.3%	-10.5%								
								-11.3%	-14.1%	-15.7%	-16.4%	-17.2%			
													-11.3%	41.6%	30.5%

ROOF TYPE: **HI**
REF COOLING: **-436**

LENGTH (m)					CHANNEL THICKNESS (m)			AIR FLOW (kg/m²)					SCHEDULE		
2	4	6	8	10	0.01	0.03	0.05	0.03	0.05	0.07	0.08	0.1	N	D	C
-18.2%	-20.6%	-22.9%	-23.8%	-24.6%											
					-24.4%	-29.9%	-21.3%								
								-22.9%	-28.9%	-32.9%	-34.2%	-36.7%			
													-22.9%	62.5%	33.8%

MADRID

The results for the climatic conditions of Madrid show possible energy savings in the order of 10–25%; the higher range with the more heavyweight roof type and further possible savings by increasing the air flow at night-time. For free-running buildings the number of hours above 25°C can be reduced by 40–55% depending on roof type, and further with an increase in air flow. The reductions in peak temperature are of about 2K.

% COOLING ENERGY SAVINGS FROM AIR RADIATORS

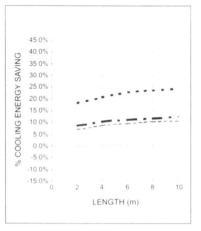

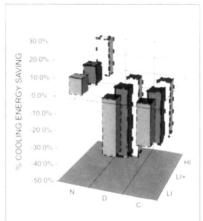

THERMAL COMFORT IMPROVEMENTS WITH AIR RADIATORS

AIR RADIATORS

NO. OF HOURS INDOORS ABOVE 25°C WITHOUT AC   Seville

ROOF TYPE: **LI**
REF COOLING: **1046**

LENGTH (m)					CHANNEL THICKNESS (m)			AIR FLOW (kg/m²)					SCHEDULE		
2	4	6	8	10	0.01	0.03	0.05	0.03	0.05	0.07	0.08	0.1	N	D	C
672	649	639	627	616											
					616	639	645								
								639	554	507	491	466			
													639	1028	738

ROOF TYPE: **LI+**
REF COOLING: **987**

LENGTH (m)					CHANNEL THICKNESS (m)			AIR FLOW (kg/m²)					SCHEDULE		
2	4	6	8	10	0.01	0.03	0.05	0.03	0.05	0.07	0.08	0.1	N	D	C
535	520	505	496	486											
					492	505	517								
								505	432	385	370	351			
													505	965	646

ROOF TYPE: **HI**
REF COOLING: **672**

LENGTH (m)					CHANNEL THICKNESS (m)			AIR FLOW (kg/m²)					SCHEDULE		
2	4	6	8	10	0.01	0.03	0.05	0.03	0.05	0.07	0.08	0.1	N	D	C
334	317	300	289	282											
					284	300	313								
								300	219	170	159	138			
													300	925	467

MADRID

PEAK INDOOR TEMPERATURE WITHOUT AC, °C   Seville

ROOF TYPE: **LI**
REF NO. HOURS: **34**

LENGTH (m)					CHANNEL THICKNESS (m)			AIR FLOW (kg/m²)					SCHEDULE		
2	4	6	8	10	0.01	0.03	0.05	0.03	0.05	0.07	0.08	0.1	N	D	C
32.4	32.3	32.3	32.3	32.3											
					32.3	32.3	34.8								
								32.3	32.1	31.9	31.9	31.7			
													32.3	33.4	32.8

ROOF TYPE: **LI+**
REF NO. HOURS: **34**

LENGTH (m)					CHANNEL THICKNESS (m)			AIR FLOW (kg/m²)					SCHEDULE		
2	4	6	8	10	0.01	0.03	0.05	0.03	0.05	0.07	0.08	0.1	N	D	C
31.5	31.5	31.4	31.4	31.4											
					31.3	31.4	31.4								
								31.4	31.1	31.0	30.9	30.8			
													31.4	32.7	32.0

ROOF TYPE: **HI**
REF NO. HOURS: **30**

LENGTH (m)					CHANNEL THICKNESS (m)			AIR FLOW (kg/m²)					SCHEDULE		
2	4	6	8	10	0.01	0.03	0.05	0.03	0.05	0.07	0.08	0.1	N	D	C
27.9	27.8	27.7	27.7	27.7											
					27.7	27.7	27.8								
								27.7	27.6	27.4	27.4	27.3			
													27.7	30.0	29.4

BIBLIOGRAPHY

NATURAL COOLING PROCESSES

Allard, F. (Ed.) (1998) *Natural Ventilation in Buildings. A Design Handbook.* James & James, London, 356 pp.

Al-Turki, A. and Zaki, G. (1991) Energy saving through intermittent evaporative roof cooling. *Energy and Buildings*, 17 (1): 35 –42.

Alvarez, S., Rodriguez, E. and Molina, J. (1991) The Avenue of Europe at Expo '92: application of cool towers. In Alvarez, S., Lopez de Asiain, J., Yannas, S. and De Oliveira Fernandes, E. (Eds) *Architecture and Urban Space.* Proceedings of the 9th PLEA International Conference, Kluwer Academic Publishers, Dordrecht, 195–201.

Argiriou, A., Santamouris, M. and Asimakopoulos, D.N. (1994) Assessment of the radiative cooling potential of a collector using hourly weather data. *Solar Energy*, 19 (8): 879–888.

Arnfield, A.J. (1979) Evaluation of empirical expressions for the estimation of hourly and daily totals of atmospheric longwave emission under all sky conditions. *Quarterly Journal of the Royal Meteorological Society*, 105: 1041–1052.

ASHRAE (1993) *Handbook of Fundamentals: SI Edition.* American Society of Heating, Refrigerating and Air-conditioning Engineers, Inc.

Bahadori, M. (1986) Natural air-conditioning systems. *Advances in Solar Energy*, 3: 310–315.

Berdahl, P. and Fromberg, R. (1982) The thermal radiance of clear skies. *Solar Energy*, 29: 299–314.

Berdahl, P. and Martin, M. (1981) Thermal radiance of skies with low clouds. *Proceedings of the International Passive and Hybrid Cooling Conference.* Miami Beach, American Section of the International Solar Energy Society, USA, 266-269.

Berdahl, P. and Martin, M. (1984) Emissivity of clear skies. *Solar Energy*, 32: 663–664.

Berger, X. (1988) A simple model for computing the spectral radiance of clear skies. *Solar Energy*, 40: 321–333.

Berger, X. and Bathiebo, J. (1989) From spectral sky emissivity to total clear sky emissivity. *Solar and Wind Technology*, 6: 551–556.

Berger, X., Bathiebo, J., Kieno, F. and Awanou, C.N. (1992) Clear sky radiation as a function of altitude. *Renewable Energy*, 2: 139–157.

Catalanoti, S., Cuomo, V., Piro, G. and Ruggi, D. (1975) The radiative cooling of selective surfaces. *Solar Energy*, 17.

Centeno, M. (1982) New formula for the equivalent night sky emissivity. *Solar Energy*, 28: 489–498.

Clark, E. (1981) Passive/hybrid comfort cooling by thermal radiation. *Proceedings of the International Passive and Hybrid Cooling Conference.* Miami Beach, American Section of the International Solar Energy Society, 682–714.

Clark, E., Loxsom, Schutt B., and Faultersack, J. (1981) Simple estimation of temperature and nocturnal heat loss for a radiantly cooled roof mass. *Proceedings of the International Passive and Hybrid Cooling Conference.* Miami Beach, American Section of the International Solar Energy Society, 244–247.

Clark, G., (1981). *A Radiative Cooling Bibliography.* Passive and Hybrid Cooling Notebook. Florida Solar Energy Association, USA.

Clark, G. and Allen, C.P. (1978) The estimation of atmospheric radiation for clear and cloudy skies. *Proceedings of the 2nd Passive Solar Conference,* 2: 676.

Clear, R.D., Gartland, L. and Winkelmann, F.C. (2003) An empirical correlation for the outside convective air-film coefficient for horizontal roofs. *Energy and Buildings,* 35 (8): 797–811.

Cole, R.J. (1979) The long wave radiation incident upon inclined surfaces. *Solar Energy*, 22: 459–462.

Conrad, G., Pytlinsky, G. and McConnel, T. (1981) Assessment of contemporary residential roof surfaces as nocturnal radiators and solar collectors. *Proceedings of the International Passive and Hybrid Cooling Conference.* Miami Beach, American Section of the International Solar Energy Society, 251–255.

Cooper, P.I., Christie, E.A. and Dunkle, R.V. (1981) A method of measuring sky temperature. *Solar Energy*, 26 (2): 153–159.

Das, A.K. and Iqbal, M. (1987) A simplified technique to compute spectral atmospheric radiation. *Solar Energy*, 39: 143–155.

Diatezua, D.M., Thiry, P.A., Dereux, A. and Caudano, R. (1996) Silicon oxynitride multilayers as spectrally selective material for passive radiative cooling applications. *Solar Energy Materials and Solar Cells*, 40: 253–259.

Duchon, C.E. and Wilk, G.E. (1994) Field comparisons of direct and component measurements of net radiation under clear skies. *Journal of Applied Meteorology*, 33 (2): 245–251.

Duffie, J.A. and Beckman, W.A. (1991*) Solar Engineering of Thermal Processes.* John Wiley & Sons, New York.

Dung Dang, PH. (1989) The cooling of water flowing over an inclined surface exposed to the night sky. *Solar and Wind Technology* 6 (1): 41–50.

Ellingson, R.G., Hai-Tien Lee, Yanuk, D. and Gruber, A. (1994) Validation of a technique for estimating outgoing longwave radiation from HIRS radiance observations. *Journal of Atmospheric and Oceanic Technology*, 11 (2), pt.1: 357–365.

Feustel, H.E. and Stetiu, C. (1995) Hydronic radiant cooling – preliminary assessment. *Energy and Buildings,* 22 (3), 193–205.

Givoni, B. (1982) Cooling by longwave radiation. *Passive Solar Journal*, 1 (3): 131–150.

Givoni, B. (1991) Performance and applicability of passive and low-energy cooling systems. *Energy and Buildings*, 17 (3): 177–199.

Givoni, B. (1994) *Passive and Low Energy Cooling of Buildings.* Van Nostrand Reinhold, New York.

Granqvist, C. (1981) *Radiative Heating and Cooling with Spectrally Selective Surfaces.* Passive and Hybrid Cooling Notebook. Florida Solar Energy Association, USA.

Guerra, J., Alvarez, S., Molina, J. and Velazquez, R. (1994) *Guia Basico para el Acondicionamiento Climatico de Espacios Abiertos.* Departamento de Ingenieria Energetica y Mecanica de Fluidos de la Universidad de Sevilla.

Harrison, A.W. (1981) Effect of atmospheric humidity on radiation cooling. *Solar Energy,* 26: 243–247.

Lechner, N. (1991) *Heating, Cooling, Lighting: Design Methods for Architects.* John Wiley & Sons, Chichester, 174–206.

Lu, S.M. and Yan W.J. (1995). Development and experimental validation of a full-scale solar desiccant enhanced radiative cooling system. *Renewable Energy,* 6 (7): 821–827.

Macho, J., Lopez, J., Felix, J., Dominguez, S. and Vila, R. (1992) *Control Climatico en Espacios Abiertos: Evaluation del Proyecto Expo '92.* Universidad de Sevilla, CIEMAT, Junta de Andalucia.

Martin, M. (1989) Radiative cooling. In Cook, J. (Ed.) *Passive Cooling.* The MIT Press, Cambridge MA, 85–137.

Meteotest (2003) *Meteonorm version 5.0.* Global meteorological database for applied climatology. Meteotest, Bern.

Oke, T. (1987) *Boundary Layer Climates.* 2nd edn. Routledge, London.

Pissimanis, D.K. and Notaridou, V.A (1981) The atmospheric radiation in Athens during the summer. *Solar Energy,* 26: 525–528.

Reaves, F. (1981) Evaporative cooling on the roof. *Proceedings of the International Passive and Hybrid Cooling Conference.* Miami Beach, American Section of the International Solar Energy Society, USA, 236–239.

Sodha, M., Kathry, A. and Malik, M. (1978) Reduction of heat flux through a roof by water film. *Solar Energy,* 20: 189–191.

Sodha, M., Ashutosh, S., Kumar, A. and Sharma, A.K. (1986) Thermal performance of an evaporatively cooled multi-storey building. *Building and Environment* 21, (2): 71–79.

Tiwari, G., Kumar, A. and Sodha, M. (1982) Cooling by water evaporation over roofs. *Energy Conversion and Management,* 22: 143–153.

Yannas, S. (2001) Passive heating and cooling design strategies. In *Climate Responsive Architecture: A Design Handbook.*

Yellott, J. (1981) *Radiative Cooling and Heating of Buildings.* Passive and Hybrid Cooling Notebook. Florida Solar Energy Association, USA.

Yellott, J. (1982) Passive and hybrid cooling systems. *Advances in Solar Energy*: 241–260.

Yellott, J.I. (1989) Evaporative cooling. In Cook, J. (Ed.) *Passive Cooling.* The MIT Press, Cambridge MA, 85–137.

ROOF DESIGN & COOLING TECHNIQUES

GENERAL

Bretz, S. and Akbari, H. (1997) Long-term performance of high-albedo roof coatings for energy efficient buildings. *Energy and Buildings,* 25 (2): 159–167.

Building Energy Analysis Group (1997) *Durability of High-Albedo Roof Coatings.* Lawrence Berkeley Laboratory.

Chandra, S., Kaushik, S. and Bansal, P. (1985) Thermal performance of a non-air-conditioned building for passive solar air-conditioning: evaluation of roof cooling systems. *Energy and Buildings,* 8: 51–69.

Colegio Oficial Arquitectos de Madrid. *Rehabilitación. La Cubierta.* Publicaciones COAM.

Dimakopoulos, I. (Ed.) (1981) *Anthology of Greek Architecture.* Ministry of Culture and Sciences, Athens.

Erell, E. and Etzion, Y. (1998) Development and Testing of an Evaporative Cooling Prototype. Roof Solutions for Natural Cooling. Final Report. European Commission Joule Programme.

Fatouros, D.A., Papadopoulos, L. and Tentokali, V. (Eds) (1979) *Studies on the Dwelling in Greece.* Paratiritis, Thessaloniki.

Feduchi, L. (1994) *Arquitectura Popular Española.* Books 1–5. Editorial Blume.

Francis, E. and Ford, B. (1999) Recent developments in passive down-draught cooling – an architectural perspective. In *European Directory of Sustainable and Energy Efficient Building.* James & James (Science) Publishers, London.

Givoni, B. (1981) Experimental studies on radiant and evaporative cooling of roofs. *Proceedings of the International Passive and Hybrid Cooling Conference.* Miami Beach, American Section of the International Solar Energy Society, USA, 279–283.

Harrison, H.W. (1997) *Roofs and Roofing: Performance, Diagnosis, Maintenance, Repair and the Avoidance of Defects.* Building Research Establishment, England.

Idris, M. (1996) Design and detailing considerations of roofs in a hot-dry climate: a case of large flat roofs. *Architectural Science Review,* 39: 193–199.

Kishore, V. (1988) Assessment of natural cooling potential for buildings in different climatic conditions. *Building and Environment,* 23 (3), 215–223.

Koenigsberger, O. and Lynn, R. (1965) *Roofs in the Warm Humid Tropics.* Architectural Association, Paper No.1. Lund Humphries, Great Britain.

Kumar, S., Tiwari, G.N. and Bhagat, N.C. (1994) Amalgamation of traditional and modern cooling techniques in a passive solar house: a design analysis. *Energy Conservation and Management*, 35 (8): 671–682.

McCampbell, H. (1991) *Problems in Roofing Design*. Butterworth Architecture.

Nayak, J. (1982) The relative performance of different approaches to the passive cooling of roofs. *Building and Environment*, 17 (2): 145–16.

Ozkan, E. (1993) Surface and inner temperature attainment of flat roof systems in hot-dry climate. *Building Research and Information*, 21 (1): 25–36.

Pearlmutter, D. (1993) Roof geometry as a determinant of thermal behaviour: a comparative study of vaulted and flat surfaces in a hot-arid zone. *Architectural Science Review*, 37: 75–86.

Pearlmutter, D., Erell., E., Etzion, Y., Meir, I. and Di, H. (1996) Refining the use of evaporation in an experimental downdraft cool tower. *Energy and Buildings*, 23: 191–197.

Philippides, D. (Ed.) (1984) *Greek Traditional Architecture*. Vols. 1–6. Melissa Publishers, Athens.

Pulín, F. (1984) Metodología de reparación de cubiertas tradicionales. *Rehabilitación de cubiertas. La cubierta. Curso de Rehabilitación*, C.O.A.M.,105–119.

Rosenfeld, A., Akbari, H., Taha, H. and Bretz, S. (1996) Implementation of light-coloured surfaces. *Proceedings of the 1996 ACEEE Summer Study on Energy Efficiency in Buildings*. Pacific Grove CA, 9: p141.

Tellez, F. and Schwartz, G. (1998) Roof Solutions for Natural Cooling. Final Report. European Commission Joule Programme.

Yannas, S. (Ed.) (1998). *Roof Solutions for Natural Cooling. Design Handbook*. European Commission Joule Programme JOR3-CT95-0074.

Yannas, S. (1999) Roof design for natural cooling. In *Proceedings. PLEA 99 Sustaining the Future*. Energy Ecology Architecture, PLEA International, 1, 427–434.

ROOF PONDS

Clark, G. (1981) *Comfort Conditions in Roof Pond Cooled Residences without Air Conditioners*. Passive and Hybrid Cooling Notebook. Florida Solar Energy Association, USA.

Cook, J. (1984) Heating and Cooling with a Roof Pond. *Award-Winning Passive Solar House Designs*, 89–93.

Crowther, K. and Meltzer, B. (1979) The thermosiphoning cool pool a natural cooling system. *Proceedings of the 3rd National Passive Solar Conference*. San Jose CA, January 11–13, 448–451.

Erell, E. and Etzion, Y. (1991) The effect of convection on a roof pond cooled by radiation in a hot-arid climate. *Proceedings of the 1991 International PLEA Conference*. Kluwer Academic Publishers, Dordrecht, 613–618.

Givoni, B. (1982) Passive indirect evaporative cooling by shaded roof ponds – a mathematical model. *Proceedings of the 5th International PLEA Conference*. Bermuda, Pergamon Press, Oxford, 131–137.

Hay, H. (1978) A passive heating and cooling system from concept to commercialisation. *Proceedings of the Annual Meeting of the American Section of the International Solar Energy Society*.

Hay, H. (1981a) *Integration of Radiative Systems*. Passive and Hybrid Cooling Notebook. Florida Solar Energy Association, USA.

Hay, H. (1981b) *Thermopond: Applicability to Climate and Structure*. Passive and Hybrid Cooling Notebook. Florida Solar Energy Association, USA.

Hay, H. (1984) 100% natural thermal control-plus. *Proceedings of the 3rd International PLEA Conference*. Mexico, Pergamon Press, Oxford.

Hay, H. (1985) Roof ponds ten years later. *Proceeding of the 10th National Passive Solar Conference*. American Solar Energy Society, USA, 181–186.

Hay, H. (1986) Roof ponds: the humidity issue. *Proceedings of the 11th National Passive Solar Conference*. American Solar Energy Society, USA, 181–186.

Hay, H. and Yellott, J. (1969) Natural air conditioning with roof pond and movable insulation. *ASHRAE Transactions*, 165–177.

Keskinel, S., Schiler, M. and Koenig, P. (1984) *A Super Insulated Mobile Home that Utilises a Roof Pond in Hot, Arid Climates*. School of Architecture, University of Southern California.

Lord, D. (1999) *An Interactice Web-based Computer Program for Thermal Design of Roof Ponds*. California Polytechnic State University.

Loxsom, F., Mosley, J. Kelly, B. and Clark, E. (1981) Measured ceiling heat transfer coefficients for a roof pond cooling system. *Proceedings of the International Passive and Hybrid Cooling Conference*. Miami Beach. American Section of the International Solar Energy Society, USA, 279–283.

Mancini, T. (1983) The performance of a roof pond solar house: The New Mexico State University experience. *Proceedings of the 8th National Passive Solar Conference*. American Solar Energy Society, USA, 811–815.

Marlatt, W., Murray, K. and Squier, S. (1984) Installation, operation, and maintenance experiences with roof pond systems. *Proceedings of the 9th National Passive Solar Conference*. American Solar Energy Society, USA, 69–74.

Marlatt, W., Murray, K. and Squier, S. (1984) *Roof Pond Systems*. Energy Technology Centre, California, USA.

Moore, F. (1993) *Environmental Control and Systems*. McGraw-Hill, USA, 195–200.

Raeissi, S. and Taheri, M. (1996) Cooling load reduction of buildings using passive roof options. *Renewable Energy*, 7 (3): 301–313.

Rodriguez, E., Molina, J.L., Guerra, J.J. and Esteban, C.J. (1998) Detailed modelling of roof ponds. ROOFSOL Project, Task 2 Final Report. European Commission Joule Programme.

Sodha, M., Singh, S. and Kumar, A. (1985) Thermal performance of a cool-pool system for passive cooling of a non-conditioned building. *Building and Environment*, 20 (4): 233–240.

Yellott, J. and Hay, H. (1969) Thermal analysis of a building with natural air conditioning. *ASHRAE Transactions*, 178–189.

COOLING RADIATORS

Addeo, A., Nicolais, L., Romeo, G., Bartoli, B., Coluzzi, B. and Silvestrini, V. (1980) Light selective structures for large scale national air conditioning. *Solar Energy*, 24: 93–98.

Berdahl, P. (1984) Radiative cooling with MgO and LiF layers. *Applied Optics*, 23 (3): 370–372.

Berdahl, P., Martin, M. and Sakkal, F. (1983) Thermal performance of radiative cooling panels. *International Journal of Heat Mass Transfer*, 26 (6): 871–880.

Berger, X. and Schneider, M. (1981) New techniques of natural cooling. *Proceedings of the International Passive and Hybrid Cooling Conference*. Miami Beach, American Section of the International Solar Energy Society, USA, 261–265.

Berger, X. *et al.* (1984) Thermal comfort in hot dry countries: radiative cooling by 'diode' roof. *Proceedings of the 3rd International PLEA Conference*. Mexico. 960–967.

Brookes, J.R. (1982) *Development of Radiative Cooling Materials*. Final Technology Report: FY 1980–1981, DOE Contract No. DE-FC03-80SF11504.

Brown, P. and Clark, G. (1984) Effect of design changes on performance of a trickle roof cooled residence. *4th Proceeding of the 9th National Passive Solar Conference*. American Solar Energy Society, USA, 51–57.

Clark, E. and Berdahl, P. (1980) Radiative cooling: Resource and applications. In Miller, H. (Ed.) *Passive Cooling Handbook*. Florida Solar Energy Association, 177–212.

Dimoudi, A., Sutherland, G., Androutsopoulos, A. and Vallindras, M. (1998) Development and Testing of a Prototype Radiative Cooling Component Using a Water Radiator Linked to a Cooling Panel. Roof Solutions for Natural Cooling. Final Report. European Commission Joule Programme.

Erell, E. and Etzion, Y. (1992) A radiative cooling system using water as a heat exchange medium. *Architectural Science Review*, 35 (2): 39–49.

Erell, E. and Etzion, Y. (1996) Heating experiments with a radiative cooling system. *Building and Environment*, 31 (6): 509–517.

Erell, E. and Etzion, Y. (1998) Radiative cooling with flat-plate solar collectors. *Building and Environment*, 35 (4): 297–305.

Etzion, Y. and Erell, E. (1989) A hybrid radiative–convective cooling system for hot-arid zones. In *Clean and Safe Energy Forever*. Proceedings, ISES Solar World Congress, Kobe, Japan.

Etzion, Y. and Erell, E. (1991) Thermal storage mass in radiative cooling systems. *Building and Environment*, 26 (4): 389–394.

Etzion, Y. and Erell, E. (1998) Low-cost long-wave radiators for passive cooling of buildings. *Architectural Science Review*.

Givoni, B. (1977) Solar heating and night radiation cooling by a Roof Radiation Trap. *Energy and Buildings*, 1: 141–145.

Givoni, B. (1982) *Cooling Buildings by Long Wave Radiation – Review and Evaluation*. Research Report to Israeli Ministry of Energy and US Dept. of Energy. J. Blaustein Institute for Desert Research, Ben-Gurion University of the Negev, Sede-Boqer Campus.

Rodriguez, E.A. and Molina, J.L. (1998) Modelling of water-based and air-cooling radiators. *ROOFSOL Project Task 2 Final Report*. European Commission Joule Programme.

PLANTED ROOFS

CIC (1997) *La cubierta de nuestro tiempo*. Cubierta ecológica Danosa, No. 302.

Eumorfopoulou, K., Ekonomides G. and Aravantinos D. (1994) Energy efficiency of planted roofs. *Proceedings of the 11th International PLEA Conference*. Desert Architecture Unit, Sede Boker, 390–397.

Harazono, Y. *et al.* (1990) Effects of rooftop vegetation using artificial substrates on the urban climate and the thermal load of buildings. *Energy and Buildings*: 435–442.

Johnston, J. and Newton, J. (1996) *Building Green: A Guide to Using Plants on Roofs, Walls and Pavements*. London Ecology Unit, Great Britain, 47–72.

Köhler, M., Schmidt, M., Grimme, F.W., Laar, M., de Assunção Paiva, V.L. and Tavares, S. (2002) Green roofs in temperate climates and in the hot-humid tropics. *Environmental Management and Health*, 13 (4): 382–391.

Kralli, M.N., Kambezidis, H.D. and Cassios, C.A. (1996) 'Green roofs' policies in cities with environmental problems. *FEB – Fresenius Environmental Bulletin*, 5: 424–429.

McPherson, G., Herrington, L. and Heisler, G. (1988) Impacts of vegetation on residential heating and cooling. *Energy and Buildings,* 12: 41–51.

McPherson, G., Simpson, J. and Livingston, M. (1989) Effects of three landscape treatments on residential energy and water use in Tucson, Arizona. *Energy and Buildings,* 13: 127–138.

Niachou, N., Papakonstantinou, K., Santamouris, M., Tsangrassoulis, A. and Mihalakakou, G. (2001) Analysis of the green roof thermal properties and investigation of its energy performance. *Energy and Buildings,* 33: 719–729.

Onmura, S., Matsumoto, M. and Hokoi, S. (2001) Study on the evaporative cooling effect of roof lawn gardens. *Energy and Buildings,* 33: 653–666.

Ostroluczky, M. *et al.* (Eds) (1998) *Green Design.* Ybl Miklos Polytechnic, Budapest.

Palomo del Barrio, E. (1998) Analysis of the cooling potential of green roofs in buildings. *Energy and Buildings,* 27: 179–193.

Theodosiou, T. (2003) Summer period analysis of the performance of a planted roof as a passive cooling technique. *Energy and Buildings,* 35: 909–917.

Wong, N., Chen, Y., Ong, C., Cheong, D. and Sia, A. (2003) Investigation of thermal benefits of rooftop garden in the tropical environment. *Building and Environment,* 38: 261–270.

Wong, N., Cheong, D., Yan, H., Soh, J., Ong, C. and Sia, A. (2003) The effects of a rooftop garden on energy consumption of a commercial building in Singapore. *Energy and Buildings,* 35: 353–364.

MATHEMATICAL MODELS

WATER PONDS

A dynamic model was developed in the context of the ROOFSOL Project (Molina and Rodriguez, 1998) to study the thermal performance of water ponds subject to various atmospheric conditions. The model has a general form that lends itself to various applications. Solution of the equations provides the pond temperature, and the water mass evaporated while the system is functioning. The model integrates all the mechanisms of heat transfer occurring in the system. The possibility that the water pond may be covered is considered. The only restriction is that the model cannot consider vertical gradients of temperature inside the water mass.

Energy Balance in a Water Pond

For shallow water ponds such as those considered here the entire water mass can be assumed to be at the same temperature. The pond temperature is then a function of time only. Under such conditions, the pond temperature, T_w, can be calculated from a global energy balance as shown in Fig. A.1.

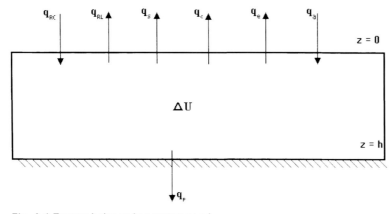

Fig. A.1 Energy balance in a water pond.

We can obtain:

where,
ΔU is the heat stored in the pond per unit pond surface area and time,
q_{RC} is the solar heat gain;
q_{RL} is the heat flux due to radiation from the pond surface;

q_c is the heat flux due to convection from the pond surface;

q_e is the heat flux due to evaporation from the pond surface;

q_s is the heat flux from sprays;

q_a is the heat flux from ambient;

q_p is the conduction heat loss to the earth surrounding the pond.

All of the fluxes are in W/m² .

Absorption of Solar Radiation

Water selectively absorbs and scatters the incident solar radiation. Comparison of data collected by several investigators has shown that all of the scattered energy propagates in the direction of the incident beam. In other words, the attenuation of incident solar radiation in water can be regarded to proceed without scattering, with the unabsorbed energy propagated in its original direction. There are numerous factors that affect the attenuation of solar radiation in water ponds. They include, among others, the spectral distribution of the optical properties of water, whether the radiation is direct or diffuse, the angle of incidence of the direct beam, the thickness of the water layer and the reflectivity of the pond floor.

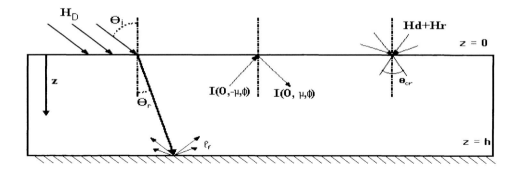

Fig. A2 Solar radiation absorption in solar ponds.

To develop the mathematical formulation for the attenuation of incident solar radiation in a water pond the following assumptions were made:

a. The air–water interface is flat so that Fresnel's equations for reflectance are applicable on both sides of the interface.
b. All the unabsorbed energy propagates in its original direction.
c. The water–air interface reflects specularly and the surface of the pond floor reflects diffusely.
d. Emissions of radiation by the water and the pond floor surface are negligible at the temperatures encountered in ponds.
e. Solar radiation incident on the water surface is assumed to consist of direct (H_D) and diffuse (H_d) components.
f. The pond is large enough to neglect the edge effects and treat the water in the pond as a one-dimensional layer of thickness **h**.

Because of the strong dependence of the absorption coefficient of water on wavelength, it is essential to divide the wavelength spectrum into a number of bands each having a uniform but different absorption coefficient. The amount of solar energy absorbed in the water mass q_{RC} and the amount of solar energy absorbed by the pond bottom q_{per} are determined.

The absorption coefficient for a band λ is given by:

$$\alpha_\lambda = \frac{-1}{I_\lambda(z,\mu,\phi)} \times \frac{\partial I_\lambda(z,\mu,\phi)}{\partial s} = \frac{-\mu}{I_\lambda(z,\mu,\phi)} \times \frac{\partial I_\lambda(z,\mu,\phi)}{\partial z}$$

(1)

Here, μ is the cosine of the angle between the direction of the radiation beam and the positive z-axis, $s = z/\mu$ and $I_\lambda(z,\mu,\phi)$ the radiation intensity. Once the radiation intensity $I_\lambda(z,\mu,\phi)$ for any given band λ is available, the net radiation flux at any location is obtained from:

$$q_{neto,\lambda}(z) = \int_0^{2\pi} \int_{-1}^1 I_\lambda(z,\mu,\phi) \times \mu \times d\mu d\phi$$

(2)

The absorption coefficient defined by eq. (1) can obtain $I_\lambda(z,\mu,\phi)$ for each band:

$$\mu \frac{\partial I_\lambda(z,\mu,\phi)}{\partial z} + \alpha_\lambda \times I_\lambda(z,\mu,\phi) = 0 \qquad \begin{cases} 0 \le z \le h \\ -1 \le \mu \le 1 \\ 0 \le \phi \le 2\pi \end{cases}$$

(3)

For the calculation of the radiation intensity $I_\lambda(z,\mu,\phi)$ a boundary condition is required at the air–water interface ($z = 0$),

$$I_\lambda(0,\mu,\phi) = \left[1 - \rho_a(\mu_i)\right] \times \frac{H_{D,\lambda}}{\mu_r} \times \delta(\phi - \phi_r) \times \delta(\mu - \mu_r) +$$

(4)

$$+ \left[1 - \rho_a(\mu')\right] \times n^2 \times \frac{H_{d,\lambda}}{\pi} + \rho_w(\mu) \times I_\lambda(0,-\mu,\phi)$$

Where $\rho_a(\mu_i)$ is the Fresnel reflectivity on the air side, μ_i is the cosine of the angle between the direction of the radiation beam and the positive z-axis, μ_r is the cosine of the angle of refraction v_r, $\rho_w(\mu)$ the Fresnel reflectivity on waterside, $H_{D\lambda}$ is the portion of the direct solar radiation in the wavelength band λ and is Hd,λ the diffuse radiation flux incident on the pond surface in the wavelength band λ. The cosine of angle of incidence μ' is related to the cosine of angle of refraction by Snell's law:

$$\mu' = \left[1 - n^2 \times \left(1 - \mu^2\right)\right]^{1/2}$$

The angles $\phi = \phi_r$ and $\mu = \mu_r$ specify the direction of the direct solar radiation beam in water.

The first term on the right-hand side of eq. (4) is the contribution of the direct radiation beam refracted in the direction $\phi = \phi_r$, $\mu = \mu_r$. The second term is the contribution of the diffuse radiation entering the water after being refracted, and the last term is the radiation from the interior of the body of water that is reflected internally at the water–air interface.

The unabsorbed portion of the radiation reaching the pond bottom will be reflected uniformly in all directions:

$$I_\lambda(h, -\mu, \phi) = \rho_f \times \frac{q_{f,\lambda}}{\pi} \quad \mu > 0 \tag{5}$$

Where ρ_f is the reflectivity of the bottom surface of the pond.

The radiation problem defined by eq. (2), (3), (4) and (5) can readily be solved. We can obtain the net radiation flux at any location, and the radiation flux reaching the pond bottom as:

$$q_{neto,\lambda}(z) = \left[(1 - \rho_a(\mu_i))\right] \times H_{D,\lambda} \times e^{\frac{-\alpha_\lambda z}{\mu_r}} + 2n^2 \times H_{d,\lambda} \int_{\mu_{cr}}^1 \left[1 - \rho_a(\mu')\right] \times \mu \times e^{\frac{-\alpha_\lambda z}{\mu}} d\mu +$$

$$+ 2\rho_f \times q_{f,\lambda} \times \left[\int_0^1 \rho_w(\mu) \times \mu \times e^{\frac{-\alpha_\lambda(h+z)}{\mu}} d\mu - \int_0^1 \mu \times e^{\frac{-\alpha_\lambda(h-z)}{\mu}} d\mu\right] \tag{6}$$

$$q_{f,\lambda} = \frac{1}{1 - 2\rho_f \times \varphi} \times \left[\left[1 - \rho_a(\mu_i)\right] \times H_{D,\lambda} \times e^{\frac{-\alpha_\lambda h}{\mu_r}} + 2n^2 \times H_{d,\lambda} \int_{\mu_{cr}}^1 \left[1 - \rho_a(\mu')\right] \times \mu \times e^{\frac{-\alpha_\lambda h}{\mu}} d\mu\right]$$

where,

$$\varphi = \int_0^1 \rho_w(\mu) \times e^{\frac{-2 \times \alpha_\lambda \times h}{\mu}} \mu \times d\mu$$

Knowing the above quantities for each wavelength band, the total quantities over the entire wavelength spectrum are obtained by superposition. The amount of solar energy absorbed in the water mass is determined from,

$$q_{RC} = \sum_{\lambda=1}^k \left[q_{neto,\lambda}(0) - q_{neto,\lambda}(h) \right] \tag{7}$$

The net radiation heat flux at $z = h$ is the amount of solar energy absorbed by the pond bottom surface and is determined from,

$$q_{per} = \sum_{\lambda=1}^k (1 - \rho_f) \times q_{f,\lambda} \tag{8}$$

where k is the total number of bands.

Heat and Mass Transfer on the Pond Surface

Radiative and convective heat transfers occur at the pond surface. Mass transfer takes place on the pond surface due to water evaporation.

Radiative heat flux

The radiative heat flux per unit pond surface area q_{RL}, can be obtained from the equation:

$$q_{RL} = \sigma \times \varepsilon_w \times (T_w^4 - T_c^4)$$

where
σ is the Stefan-Boltzmann constant,
ε_w is the water pond emissivity,
T_c is the effective temperature of sky and T the water temperature (both are expressed in Kelvin).

The emissivity of water was taken as 0.95. We considered the sky as a black body at an effective temperature T_c and that the water pond is not seen by any other surfaces.

Convective heat flux

The convective heat transfer between the pond surface and the surrounding air can be expressed as:

$$q_c = h_c \times (T_w - T_a)$$

where,

h_c is the heat transfer coefficient ($W/m^2 °C$),

T_w (°C) is the temperature of the pond surface and T_a is the ambient dry bulb temperature (°C).

In general, the convective heat transfer is due to forced convection, but at low wind speeds free convection will prevail. Convection is taken into account in the calculation of the heat transfer coefficient.

Evaporative heat flux

The amount of heat lost due to water being evaporated from the pond surface is given in terms of the mass of water evaporated and the latent heat hl_{ig},

$$q_e = \frac{K_m}{R \times T_a} \times h_{lg} \times \left[P_{vs}(T_w) - HR \times P_{vs}(T_a) \right]$$

where,
K_m is the mass transfer coefficient calculated using empirical correlations,
T_a is the air temperature in K;
P_{vs} is obtained as the water saturation pressure at water pond temperature (T) or air temperature,
HR is the relative humidity.

Heat Loss From Operation of Water Sprays

The amount of water evaporated in a spray can be estimated from a simple model for an isolated drop which evaporates in a mass of air. Using such a model it is possible to obtain the final drop radius (r_f) and the final drop temperature (T_f). If $m9$ is the mass flow circulated through the sprayers, the amount of evaporated water is given by Molina and Rodriguez (1998):

$$m'_{evap} = m' \times \left(1 - \frac{r_f^3}{r_i^3} \right)$$

and, hence, the heat lost by water evaporation would be:

$$q_s = m' \times \rho_1 \times c_p \times \left(T_i - \frac{r_f^3}{r_i^3} \times T_f \right)$$

Heat Input to Pond from the Addition of Water

The heat input to the pond can be calculated by:

$$q_a = m' \times c_p \times T_{apor}$$

where,
$m9$ is the mass flow of water,
C_p is the specific heat of water,
T_{apor} is the temperature of the water entering the system.

Pond Cover

It is assumed that heat transfer between the pond and the ambient air is reduced by use of a cover on the pond. For this, we have developed a model of the covering and coupled this with the system. The pond cover and the air layer between the cover and the pond are modelled as two conduction layers in steady state.

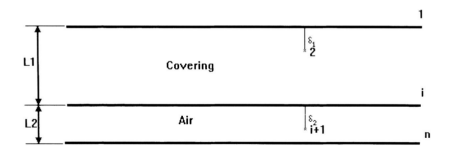

Fig. A.3 Pond covering.

The convective heat flux, q_c, is calculated by imposing the contour temperatures. We can write the solution in the form:

$$q_c = \frac{T_c - T_w}{\dfrac{L_1}{k_c} + \dfrac{L_2}{k_a}}$$

where,
q_c, is the convective heat flux in W/m²;
T_c is the temperature of the cover, which can be calculated taking into account the radiation impinging on it and the external air temperature;
T_w is the water temperature;

L_1 is the thickness of the cover;

L_2 is the thickness of the layer of air between the cover and the water;

k_c is the thermal conductivity of the cover and k_a is the thermal conductivity of the air.

Coupling of the pond cover with the system

When we couple the cover with the pond system, we find that the equations that determine the system model change. These changes are considered here. The equations that must be employed are:

a. The boundary condition at the upper surface of the cover is:

$$q_c = \alpha_c \times q_{RC} + h_{rad} \times (T_{sky} - T_c) + h_c \times (T_a - T_c)$$

where,

q_c is the heat flux;

q_{RC} is the solar radiation;

α_c is the absorptivity of the upper surface of the covering;

q_c is the radiative heat transfer coefficient;

h_{rad} is the sky temperature;

T_c is the temperature on the upper surface of the covering;

h_c is the convective heat transfer coefficient;

T_a is the air temperature.

b. Balance in pond. Owing to the presence of the cover this is reduced to:

$$\Delta U = \frac{\rho \times c_p \times V}{A} \times \frac{dT}{dt} = -q_c - q_p$$

where,

q_c is the conduction heat loss from pond to air through the covering;

q_p is the conduction heat loss to the earth surrounding the pond.

With these equations, we can calculate the thermal performance of the water pond, when the pond is covered.

Conclusion

The model presented here calculates the water pond temperature and the water mass evaporated while the system is functioning. The model is limited to systems without vertical temperature gradients inside the water mass. The model can be easily coupled to a model for the thermal behaviour of the complete building.

WATER-BASED RADIATORS

The modelling of water-based radiators as applied for cooling purposes does not differ much from that of water-based solar collectors. In fact, commercial flat-plate solar collectors can be used as radiators for cooling by removing their transparent cover so that the longwave radiant cooling mechanism is not inhibited. However, since the area needed to produce a significant cooling effect by these radiators is larger than that for solar heating, it is usually more convenient to build a radiator for the specific application of cooling, instead of modifying commercial solar panels.

For modelling water-based radiators we have assumed the typical plate and tube geometry with back insulation and no cover. The plate can be finished with a selective coating, thus we will consider that possibility in the model by incorporating three emissivities (ε_1, ε_2 and ε_3) corresponding to three emission bands: that of the so-called atmospheric window (8–13 μm) and two adjacent bands (of 0.1–8 μm and 13–100 μm respectively). Approximately, the radiative fractions for these three bands are F_{r2} = 32% for the atmospheric window and F_{r1} = 14% and F_{r3} = 54% for the other two bands.

The radiator is composed of a plate of thickness e_p, thermal conductivity k_p, a back thermal insulation with thickness e_a and conductivity k_a. The plate radiator is attached to a set of parallel pipes separated a distance w from each other and with external and internal diameters D_e and D_i, respectively. The material of the pipes has a thermal conductivity of k_t. The pipes are attached to the plate with a good contact. The bond resistance between them is R_b. Recommended values for this resistance are below 0.030 m² K/W. The mass flow rate of water per metre of plate width is m_w and the total length of the radiator L. The following equations show the calculations needed to obtain the performance of a water-based radiator.

Calculation of the convective heat transfer coefficient (h_{cv}) between the plate and the ambient temperature. This coefficient depends mainly on the wind velocity (v):

$$h_{cv} = 5.7 + 3.8v \qquad \text{if } v \leq 4 \text{ m/s}$$

$$h_{cv} = 7.3v^{0.8} \qquad \text{if } v > 4 \text{ m/s}$$

Calculation of the linearised radiant heat transfer coefficients. There is one average (h_r) for the whole spectrum and one for the atmospheric window (h_{r2}):

$$h_r = 4\sigma T_m^3(\varepsilon_1 F_{r1} + \varepsilon_2 F_{r2} + \varepsilon_3 F_{r3})$$

$$h_{r2} = 4\sigma T_m^3\varepsilon_2 F_{r2}$$

where σ is the Stefan–Boltzmann constant (σ=5.67e-8 W/m²/K⁴) and T_m is an average radiant exchange temperature in Kelvin that can be supposed constant for all radiant exchanges with little error.
Calculation of the global heat transfer coefficient (U_L) between plate and ambient:

$$U_L = h_{cv} + h_r + \frac{1}{e_a/k_a + 1/(h_{cv} + h_r)}$$

Calculation of the fin efficiency (F) of the plate between tubes:

$$F = \frac{\tan h(ml)}{ml}$$

where,

$$ml = \sqrt{\frac{U_{\mathrm{L}}}{k_{\mathrm{p}} e_{\mathrm{p}}}} \frac{(w - D_{\mathrm{e}})}{2}$$

Calculation of the interior heat transfer coefficient in the pipes (water-tube) (h_{fi}). There are two methods for calculating h_{fi} depending of the regime of the water flow (laminar or turbulent). The basic expression is:

$$h_{\mathrm{fi}} = Nu \frac{k_{\mathrm{w}}}{D_{\mathrm{i}}}$$

where k_{w} is the water thermal conductivity, Nu is the Nusselt number which, for laminar flow, is equal to 3.66 and for turbulent flow will be calculated from the following expression:

$$\mathrm{Nu} = 0.023 Re^{0.8} Pr^{0.333}$$

where Re is the Reynolds number calculated from the water velocity inside the tubes (v_{t}), the interior diameter of them (D_{i}) and the kinematic viscosity of the water (v). Pr is the Prandtl number of the water:

$$\mathrm{Re} = \frac{v_{\mathrm{t}} D_{\mathrm{i}}}{v}$$

Calculation of the collector efficiency factor (F'):

$$F' = \frac{1/U_{\mathrm{L}}}{w \left[\dfrac{1}{U_{\mathrm{L}} \left[(w - D_{\mathrm{e}})F + D_{\mathrm{e}} \right]} + R_{\mathrm{b}} + \dfrac{\log(D_{\mathrm{e}}/D_{\mathrm{i}})}{2\pi k_{\mathrm{t}}} + \dfrac{1}{h_{\mathrm{fi}} \pi D_{\mathrm{i}}} \right]}$$

Calculation of the collector heat removal factor (F_{R}):

$$F_{\mathrm{R}} = \frac{m_{\mathrm{w}} C_{\mathrm{w}}}{n w L U_{\mathrm{L}}} \left[1 - \exp\left(\frac{-n w L F' U_{\mathrm{L}}}{m_{\mathrm{w}} C_{\mathrm{w}}} \right) \right]$$

where n is the number of tubes per metre of radiator width and C_{w} is the specific heat of the water.

The heat flux evacuated by each metre of radiator width (Q_{rad}) is given by this expression:

$$Q_{\mathrm{rad}} = n w L F_{\mathrm{R}} \left[U_{\mathrm{L}} (T_{\mathrm{wi}} - T_{\mathrm{a}}) + \frac{h_{\mathrm{r2}} (T_{\mathrm{a}} - T_{\mathrm{sk}})}{F_{\mathrm{r2}}} \right]$$

where T_{a} is the ambient temperature, T_{wi} is the inlet water temperature of the radiator and T_{sk} is the average sky effective temperature. In this expression, the first term inside the brackets represents the heat flux evacuated by the radiator plate due to convection to ambient and due to radiation to ambient temperature (radiation to the sky in the spectral bands 1 and 3). The second term is the radiant contribution of the atmospheric window to the heat flux evacuated by the radiator.

From the heat flux expressed above the water outlet temperature (T_{wo}) of the radiator can be easily derived:

$$T_{\mathrm{wo}} = T_{\mathrm{wi}} - \frac{Q_{\mathrm{rad}}}{m_{\mathrm{w}} C_{\mathrm{w}}}$$

AIR-BASED RADIATORS

Air-based radiators perform less well than water radiators of the same area. This is because the heat transfer coefficients and thermal capacity of the air are sensibly lower than those of water and performance is very much conditioned by these factors. Storage of the coolness produced at night by air radiators requires a separate medium whereas with water radiators storage can be in water tanks, roof ponds or in cooling panels.

The simplest air radiator is composed of a metal plate and a back insulation separated by a certain distance (e) configuring an air channel inside which the air flows. The plate cools by radiation and convection and transmits this cooling effect to the air in the channel by convection. As with water radiators, the plate can be finished with a selective coating on its outer surface, thus we will consider that possibility in the model by assuming three emissivities (ε_1, ε_2 and ε_3) corresponding to the three emission bands. These three bands are the so-called atmospheric window (8–13 μm) and the two adjacent radiative bands (0.1–8 μm and 13–100 μm). Approximately, the radiative fractions for these three bands are F_{r2} = 32% for the atmospheric window and F_{r1} = 14% and F_{r3} = 54% for the other two bands.

In this case the plate will behave more uniformly because the fluid (air) is in contact with all of the plate and there are no preferential paths (as was the case with the tubes in a water radiator). Therefore the fin effect of the plate will not be considered here. We neglect the thermal resistance of the plate, assuming that it is sufficiently conductive. For the back insulation we assume perfect insulation (back heat flow equal to zero). The mass flow rate of air per metre of plate width is m_a and the total length of the radiator will be called L.

The following equations show the calculations needed to obtain the performance of air-based radiators. The calculation of the convective heat transfer coefficient (h_{cve}) between the plate and the ambient air is the same as for water radiators, and the same equations are reproduced here. This coefficient depends mainly on the wind velocity (v):

$$h_{cve} = 5.7 + 3.8v \qquad \text{if } v \leq 4 \text{ m/s}$$

$$h_{cve} = 7.3v^{0.8} \qquad \text{if } v > 4 \text{ m/s}$$

The same equations apply to the calculation of the linearised radiant heat transfer coefficients. There is an average coefficient (h_{re}) for the whole spectrum and one for the atmospheric window (h_{re2}):

$$h_{re} = 4\sigma T_m^3 (\varepsilon_1 F_{r1} + \varepsilon_2 F_{r2} + \varepsilon_3 F_{r3})$$

$$h_{re2} = 4\sigma T_m^3 \varepsilon_2 F_{r2}$$

where,
σ is the Stefan–Boltzmann constant (σ =5.67e-8 W/m²K⁴)
T_m is an average radiant exchange temperature in Kelvin that can be assumed to remain constant for all radiant exchanges without much error.

In the interior of the channel we have convection between the air flowing inside and the two limiting surfaces (the plate and the back surface) and longwave radiation between those two surfaces. The process of calculating the interior convective heat transfer coefficient in the channel (h_{fi}), is similar to that illustrated for the tubes in water radiators. The only difference is the characteristic length which in this case is the hydraulic diameter (D_h) of a rectangular channel:

$$D_h = \frac{2ew}{e+w}$$

where w is the width of the radiator.

Since $w \gg e$, then D_h is approximately equal to twice the channel thickness. Again there will be two ways for calculating h_{fi} depending of the regime of the air flow (laminar or turbulent). The basic expression is:

$$h_{fi} = \text{Nu}\frac{k_a}{D_h}$$

where k_a is the air thermal conductivity, Nu is the Nusselt number which for laminar flow is equal to 3.66 and for turbulent flow will be calculated from the following expression:

$$\text{Nu} = 0.020\,\text{Re}^{0.8}$$

where Re is the Reynolds number calculated from the air velocity inside the channel (v_{ch}), its hydraulic diameter (D_h) and the kinematic viscosity of the air (v). In the above equation a Prandtl number of the air equal to 0.7 has been assumed:

$$\text{Re} = \frac{v_{ch}D_h}{v}$$

The radiant exchange between the plate and the back surface of the channel is linearised through an interior radiant heat transfer coefficient (h_{ri}) calculated as follows:

$$h_{ri} = \frac{4\sigma T_m^3}{2/\varepsilon - 1}$$

where T_m is an average radiant exchange temperature in Kelvin that can be assumed constant for all radiant exchanges without much error. For simplicity the same emissivity (ε) has been assumed for the two surfaces of the channel. Later we will demonstrate that its effect in the radiator performance is not very relevant.

Two useful heat transfer coefficients will be calculated, one is related to the outside convective and radiant exchanges (U_e) and the other is related to the same exchanges inside the channel (U_i):

$$U_e = h_{re} + h_{cve}$$

$$U_i = h_{fi} + \frac{h_{ri}h_{fi}}{h_{ri} + h_{fi}}$$

The dimensionless expression for calculating the outlet temperature (T_{fo}) of the air from a radiator of length L is:

$$\frac{T_{fo} - T_a + \dfrac{h_{re2}(T_a - T_{sk})}{h_{re}U_e}}{T_{fi} - T_a + \dfrac{h_{re2}(T_a - T_{sk})}{h_{re}U_e}} = \exp\left[-\frac{U_e U_i L}{(U_e + U_i)m_a C_a}\right]$$

where T_{fi} is the inlet air temperature, T_a is the ambient temperature and T_{sk} is the average sky effective temperature.

The heat flux evacuated by each metre of radiator width (Q_{rad}) is given by this expression:

$$Q_{rad} = m_a C_a (T_{fi} - T_{fo})$$

COOLING PANELS

Cooling panels consist of a group of tubes embedded in a concrete slab placed on the ceiling of a room. A cold fluid is pumped through the tubes and a fraction of the cooling load of the room is satisfied by this system. The present study develops a two-dimensional transient numerical model of the cooling panel that can be converted easily, by programming in a model for a complete cooling panel. The numerical model is capable of predicting both steady state temperature profiles, transient responses and also the evolution with time of heat flow per unit of length.

Statement of the Problem

The present study solves the two-dimensional transient heat transfer conduction with constant properties, under different types of boundary conditions.

$$\frac{\partial^2 T}{\partial x^2} + \frac{\partial^2 T}{\partial y^2} = \frac{1}{\alpha}\frac{\partial T}{\partial t} \tag{1}$$

To solve eq. (1) for the temperature distribution $T(x,y,t)$ it is necessary to specify an initial condition and boundary conditions for each dimensional coordinate. The boundary and initial conditions considered in the model are:

$$t = 0 \quad T(x,y,0) = T(x,y)$$

$$x = 0 \quad y \in \left[(0,h_1) \cup (h_1 + R, H)\right] \Rightarrow \quad -k\frac{\partial T}{\partial x} = 0 \qquad \text{(a)}$$

$$x = L \quad y \in (0,H) \qquad\qquad \Rightarrow \quad -k\frac{\partial T}{\partial x} = 0 \qquad \text{(b)}$$

$$y = 0 \quad x \in (0,L) \qquad\qquad \Rightarrow \quad -k\frac{\partial T}{\partial y} = h_b\,(T - T_{rb})$$

$$y = H \quad x \in (0,L) \qquad\qquad \Rightarrow \quad -k\frac{\partial T}{\partial y} = h_a\,(T - T_{ra})$$

$$x^2 + (y - h_1)^2 = R^2 \qquad\qquad \Rightarrow \quad -k\frac{\partial T}{\partial n} = h_t\,(T - T_t)$$

where,

h_a, h_b are the heat transfer coefficients of the upper and lower surface of the panel, W/m²/K;

h_t is the heat transfer coefficient of the inside surface of the tube, W/m²/K;

T_{ra}, T_{rb}, T_{ra} are the room air temperatures above the upper and below the lower surface of the panel and the local fluid temperature inside the tube, °C;
k is the thermal conductivity of the material, W/m/K,

The boundary conditions (a) and (b) are due to the symmetry conditions of the problem.

Finite Element Model

The Galerkin method

Conduction is treated by means of a Finite Element Model:

$$\rho C_p \frac{\partial T}{\partial t} = k \frac{\partial^2 T}{\partial x^2} + k \frac{\partial^2 T}{\partial y^2} + G \quad \text{over the region A} \tag{2}$$

$$-k \frac{\partial T}{\partial x} nx - k \frac{\partial T}{\partial y} ny = q + h(T - Ta) \quad \text{over the boundary } S_B \tag{3}$$

$$T = Tc \quad \text{over the boundary } S_A \tag{4}$$

When we reduce a continuum problem (2) to a finite set of unknowns, we must define by e_A and e_S, respectively, the error involved in satisfying eqs (2) and (3) by a discretisation aproximation, so that:

$$e_A = -\rho C_p \frac{\partial T}{\partial t} = k \frac{\partial^2 T}{\partial x^2} + k \frac{\partial^2 T}{\partial y^2} + G \tag{5}$$

$$e_s = k \frac{\partial T}{\partial x} nx + k \frac{\partial T}{\partial y} ny + q + h(T - Ta) \tag{6}$$

Since boundary condition type (4) is directly satisfied by the prescription of T there is no error associated with this term.
In a weighted residual approach we seek to minimize, in some global manner, the residuals e_A and e_S, weighted by suitable weighting functions W_A and W_B. So:

$$\int_A e_A W_A d_A \int_{SB} e_A W_S d_S = 0 \tag{7}$$

Substituting from (5) and (6) in (7) and applying Green's Theorem and assuming $W_S = -W_A = W$, results in:

$$-\int_A (\rho C_p \frac{\partial T}{\partial t} W + k \frac{\partial T}{\partial x} \frac{\partial W}{\partial x} + k \frac{\partial T}{\partial y} \frac{\partial W}{\partial y} - WG) d_A +$$

$$+\int_{SA} (k \frac{\partial T}{\partial x} nx + k \frac{\partial T}{\partial y} ny) W d_s - \int_{SB} [q + h(T - Ta)] W ds = 0 \tag{8}$$

where, S_A is that part of the boundary with conditions type (4). As mentioned above, the contribution to the error of this part is null, and this term can be eliminated from (8).

Finite element discretisation

If we now adopt a finite element discretisation, then T may be approximated as

$$T = \sum_{i=1}^{n} N_i T_i \qquad (9)$$

in which n is the total number of nodes, T_i are the nodal values of T and N_i are the global shape functions.
The weighting function W_i corresponding to node i can then be conveniently chosen so that $W_i = N_i$

Substituting from (9) in (8) gives

$$\sum_j \int_A N_i \rho C_p N_j d_A \frac{\partial T_j}{\partial t} + \sum_j \int_A (K \frac{\partial N_i}{\partial x} \frac{\partial N_j}{\partial x} + k \frac{\partial N_i}{\partial y} \frac{\partial N_j}{\partial y}) d_A T_j = \int_A G N_i d_A$$

$$- \int_{SB} (q - hTa) N_i d_A - \sum_j \int_{SB} N_i h N_j d_s$$

$$\sum_j m_{ij} \frac{\partial T_j}{\partial t} + \sum_j K_{ij} T_j + \sum_j H_{ij} T_j = bg_i + bh_i + bq_i$$

$$[M] \frac{\partial T}{\partial t} + ([K]) + [H] \overline{T} = \overline{B} \qquad (10)$$

where,
[M] 'Mass' matrix is the capacity matrix;
[K] 'Stiffness' matrix. In heat transfer problems it becomes the conductivity matrix;
[H] Part of the stiffness matrix that depends on the film coefficient;
B Load vector;
T Nodes temperature. Unknown vector.

Element matrix and applied load vector

Triangular elements have been used in this finite element formulation with three degrees of freedom corresponding to the value of temperature at each node.
It is assumed that T varies linearly between nodes and therefore T at any position can be represented as

$$T^{(e)} = C_1 + C_2 \times C_3 y \qquad (11)$$

in which the superscript (e) denotes the particular element under consideration.
Inserting the values of T_i and its coordinates into eq. (11) for $i = 1,2,3$ we obtain the element shape functions

$$N_i^e = \frac{1}{2A^e}(\alpha_i + \beta_i x + \gamma_i y)$$

see Hinton and Owen (1979).

The contribution to the global matrix and global load vector from each element are evaluated by integration of the expressions obtained in eq. (10) over each element. Then these elemental matrixes must be assembled into the global matrix and load/vector at the appropriate positions.

$$k_{ij}^e = \frac{k}{4A^e}(\beta_i \beta_j + \gamma_i \gamma_j)$$

$$M_{ij}^e = \rho C_p \frac{A^c}{3} \quad i = j \; ; \quad M_{ij}^e = 0 \quad i \neq j$$

M_{if}^e is a diagonal expression for the mass matrix and it offers accurate results (see Zienkiewicz, 1980).

$$H_{ij} = \frac{1}{6}hA^e s_{ij}\begin{bmatrix} 2 & 1 \\ & \\ 1 & 2 \end{bmatrix}; \qquad bg^e = \frac{1}{3}GA^e \begin{bmatrix} 1 \\ 1 \\ 1 \end{bmatrix} \qquad (\text{rows } i,j,k)$$

$$bh^e = \frac{1}{2}hTa\, s_{ij}\begin{bmatrix} 1 \\ 1 \end{bmatrix} \qquad (\text{rows } i,j); \qquad bq^e = -\frac{1}{2}q\, s_{ij}\begin{bmatrix} 1 \\ 1 \end{bmatrix} \qquad (\text{rows } i,j)$$

Transient 2D Conduction

Description of the problem and resolution method

The transient 2D problem involved the resolution of eq. (1).
This problem has been solved using the finite element method with triangular elements, which gives the following linear matrix formulation for eq. (1):

$$\begin{cases} [m]\dfrac{\partial \vec{T}}{\partial t} + ([K]+[H])\vec{T} = \vec{B} \\ \\ t = 0, \vec{T}(0) = \vec{T}_o \end{cases}$$

$$(12)$$

that can be expressed as:

$$\begin{cases} \dfrac{\partial \vec{T}}{\partial t} = [A]\vec{T} + \vec{P} \\[2em] \vec{T}(0) = \vec{T}_0 \end{cases} \tag{13}$$

where, $A = [M]^{-1}([K]+[H])$

$$P = [M]^{-1}[B]$$

The previous equation (13), is a system of ordinary differential equations with constant coefficients. This problem can be solved using the eigenvalues solution method which, considerating P constant with time, gives:

$$\vec{T} = ([V][e^{\lambda_n t}][V]^{-1})\,\vec{T}_0 + ([V]\left[\frac{1}{\lambda_{ii}}(e^{\lambda_n t}-1)\right][V]^{-1})\,\vec{P} \tag{14}$$

where V eigenvectors matrix λ_i: Eigenvalues of the problem. $\left[e^{\lambda_n t}\right]$: Diagonal matrix.

$\left[\dfrac{1}{\lambda_{ii}}(e^{\lambda_n t}-1)\right]$: Diagonal matrix.

Knowledge of T implies the previous calculation of eigenvalues (λ_i) and eigenvectors (V).

Transient 2D+1 Problem

Description of the problem and assumptions

The transient 2D+1 problem has been developed joining two submodels: "pipe" and "section".

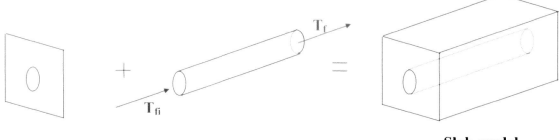

Section submodel   **Pipe submodel**   **Slab model**

A simple way of joining the submodels is the following: An energy balance is made between two sections connected by a pipe. It is assumed that the heat flow is constant along the tube and equal to that obtained in the section.

$$Q_{\text{tube}_i} L_{i,i+1} = \dot{V}_w \rho_w Cp_w (T_{w_{i+1}} - T_{w_i})$$

and therefore

$$T_{w_{i+1}} = T_{w_i} + \frac{Q_{\text{tube}_i} L_{i,i+1}}{\dot{V}_w \rho_w Cp_w}$$

where,
Q_{tube_i} is heat flow through the tube surface at section i;
$L_{i,i+1}$ is length of the pipe between sections i and $i + 1$;
V_w is water volumetric flow rate;
ρ_w is water density;
Cp_w is water heat capacity;
Tw_i is water temperature at section i;
Tw_i: water temperature at section $i + 1$.

Conclusion

The model that was developed couples the transient conduction in a slab with the embedded pipe and the slab boundaries. The finite element method was used to solve the numerical model. The results of this study have demonstrated that the cooling panel model predicts accurate heat flows and temperature fields. The results are sensitive to mesh size.
Acknowledgements
Molina and Rodriguez (1998).

PLANTED ROOFS

Introduction

This is an abridged version of a mathematical model developed in the context of the ROOFSOL Project (Palomo Del Barrio, 1998a,b).

Three components can be distinguised in a planted roof: the canopy (leaf cover), the soil and the structural support (Fig. A4 1). The quantities defining the outdoor conditions are the solar radiation flux, the thermal radiation flux from the sky, the temperature and moisture content of the air, the wind speed and the wind direction. Indoor conditions are usually defined by the thermal state of the indoor air and the temperatures of building surfaces in contact with the underside of the roof. The roof is assumed to be large enough to have horizontal homogeneity. Heat and mass fluxes are assumed to be mainly vertical, so that one-dimensional models can be used to describe the thermal behaviour of the roof components.

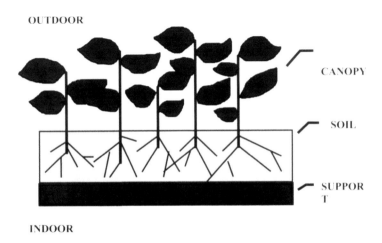

OUTDOOR

CANOPY

SOIL

SUPPORT

INDOOR

Fig. A.4. Components of a planted roof

Roof Support Model

The roof support is assumed to be an homogeneous layer of solid material with constant thermophysical properties. The energy balance equation governing the evolution of its thermal state is, in one dimension:

$$\rho c_p \, \frac{\partial \, T_s(z,t)}{\partial \, t} = \lambda_s \, \frac{\partial^2 \, T_s(z,t)}{\partial \, t^2} \qquad (1)$$

where $T_s(z,t)$ represents the temperature field, ρ is the density, c_p the specific heat, and λs the thermal conductivity of the material. The following boundary conditions, decoupling the support from the rest of the roof, are assumed:

$$\begin{cases} T_s(z=0,t) = T_{support,top}(t) \\[6ex] -\lambda_s \left. \dfrac{\partial T_s(z,t)}{\partial z} \right|_{z=L} = h\,(T_{support,bottom} - T_{in}) \end{cases} \tag{2}$$

where $z = 0$ means the top of the support layer and $z = L$ the bottom in contact with the indoor air. L is the thickness of the support layer, T_{in} the indoor air temperature, $T_{support,top}$ the imposed temperature at the top surface, and $T_{support,bottom}$ the temperature of the support at its bottom surface.

Soil Model

A soil is a porous medium in which three phases can be distinguished: the solid matrix (minerals and organic material), the liquid phase (water) and the gazeous phase (air and water vapour). Qualitatively we can say that in unsaturated soils, heat will be transported in these three phases. The respective dominant mechanisms will be: conduction in solids and liquid phases, convection in liquid and gazeous phases and latent heat transfer by vapour diffusion in pores. Furthermore, the heat transfer will depend on water content and on the temperature. This leads to a mutual dependence and continuous redistribution of heat and moisture which is quantitatively described by a set of coupled non-linear equations. The formal theory of these phenomena was first stated by Philip and de Vries (1957), de Vries (1958), and Luikov (1966) in the 1950s and 1960s. Since then, a large amount of theoretical and experimental work has been performed. The porous medium is assumed equivalent to a fictitious continuum, and the macroscopic balance equations are obtained by averaging microscopic balance and transfer equations (at the pore scale) over a representative volume (Whitaker, 1977). The following assumptions are made:

- thermal and moisture content gradients in horizontal directions are assumed to be zero (one-dimensional model);
- the soil matrix is assumed equivalent to a homogenous and isotropic continuum. Its properties do not explicitly depend on the direction;
- the liquid and vapour phases are always in equilibrium;
- pores are assumed small. The total pressure is then regarded as a constant and the net air transfer is assumed to be zero.

The macroscopic balance equations can be written as:

$$\rho c_{\mathrm{p}}(\omega,T)\,\frac{\partial\,T(z,t)}{\partial\,t}\;=\;\frac{\partial}{\partial\,z}\left\{\left[\lambda(\omega,T)\,+\,\Lambda\,D_{\mathrm{vT}}(\omega,T)\right]\frac{\partial\,T(z,t)}{\partial\,z}\,+\,\Lambda D_{\mathrm{v}\omega}(\omega,T)\,\frac{\partial\,\omega(z,t)}{\partial\,z}\right\}$$

(3)

$$\frac{\partial\,\omega(z,t)}{\partial\,t}=\frac{\partial}{\partial\,z}\left\{D_{\omega}(\omega,T)\,\frac{\partial\,\omega(z,t)}{\partial\,z}\,+\,D_{\mathrm{T}}(\omega,T)\,\frac{\partial\,T(z,t)}{\partial\,z}\right\}\,-\,\frac{\partial\,K(z,t)}{\partial\,z}\,+\,\varphi(z,t)$$

where,

$T(z,t)$ is the local temperature in the porous medium domaine (°C);
$\varpi(z,t)$ is the local volumetric moisture content in the porous medium domaine (–);
$\rho c_{\mathrm{p}}(\varpi,T)$ is the weighted heat capacity (J/kgK);
$\lambda(\varpi,T)$ is the effective thermal conductivity (Wm/K);
Λ is the latent heat of vaporization (J/kg);
$D_{\mathrm{vT}}(\varpi,T)$ is the non-isothermal vapour diffusivity coefficient (kg/m²/s/K);
$D_{\mathrm{v}\omega}(\varpi,T)$ is the isothermal vapour diffusivity coefficient (kg/m²/s);
$D_{\omega}(\varpi,T)$ is the isothermal mass (vapour+liquid) diffusivity coefficient (m²/s);
$D_{\mathrm{T}}(\varpi,T)$ is the non-isothermal mass (vapour+liquid) diffusivity coefficient (m²/s/K);
$K(z,t)$ is the hydraulic conductivity (*m/s*);
$\varphi(z,t)$ is the water sink representing the root extraction term (*s*).

Here we will assume the moisture profiles within the soil to be constant. Accordingly, eq. (3) becomes:

$$\rho c_{\mathrm{p}}(\omega,T)\,\frac{\partial\,T(z,t)}{\partial\,t}\;=\;\frac{\partial}{\partial\,z}\left\{\left[\lambda(\omega,T)\,+\,\Lambda\,D_{\mathrm{vT}}(\omega,T)\right]\frac{\partial\,T(z,t)}{\partial\,z}\right\}$$

(4)

All the coefficients in this equations are dependent variables of the moisture content and the temperature. They must be specifically determined for each soil. However, some functional relations between these coefficients and the soil temperature and moisture can be proposed. The following correlation model for estimating soil thermal conductivity was proposed in Verschinin *et al.*, (1966)

$$\lambda(\omega)\,10^{7}\;=\;\left[2.1\left(\frac{\rho}{1000}\right)^{(1.2-2\omega)}e^{-0.7(\omega-0.2)^{2}}\,+\,\left(\frac{\rho}{1000}\right)^{(0.8+2\omega)}\right]\rho c_{\mathrm{p}}(\omega)$$

(5)

where ρ is the soil apparent density (kg/m³), ϖ the volumetric moisture content, λ the thermal conductivity (J/m/s/K) and ρc_{p} the thermal capacity (J/m/³/K). A large sample of soils, with apparent density values from 1100 to 1500 (kg/m) and volumetric moisture contents from 4% to 25%, has been used in Verschinin *et al.*, (1966) to obtain the previous correlation. λ values calculated with this formula involved a relative error lower than 7% in most cases. To be coherent with the correlation model (5), the following formula, proposed by the same author, will be used for estimating the soil thermal capacity. As in the previous case, variables must be given in the International Unit System.

$$\rho c_{\mathrm{p}}(\omega) = 4180 \, (0.2 + \omega) \, \rho \tag{6}$$

It can be demonstrated that under the following assumptions:

- the air and water vapour follow the ideal gas law;
- the net transfer of air in the soil is negligible;
- pores are small enough for the total pressure to be regarded as constant;
- the soil moisture content is always above the wilting point[1] and the total water potential in equilibrium with vapour is then greater than $-10^{4.2}$ cm.

The non-isothermal soil vapour diffusivity D_{vT} (kg/m²/s/K), can be written as:

$$D_{\mathrm{vT}} = - \frac{D \, \Lambda}{R_{\mathrm{v}}^{2} \, T^{3}} \, \frac{P \, p_{\mathrm{v}}}{P - p_{\mathrm{v}}} \tag{7}$$

where,
D is the diffusion coefficient of water vapour in air (m²/s);
Λ is the latent heat of vaporization (J/kg);
R_{v} the gas constant of water vapour (J/kg/K);
T the absolute temperature (K);
P the total pressure (Pa) and pv the partial vapour pressure (Pa).

The parameter ρ_{v} is given by the following thermodynamic relation:

$$p_{\mathrm{v}} = p_{\mathrm{s}} \, \exp \left(\frac{g \, \psi}{R_{\mathrm{v}} \, T} \right) \tag{8}$$

with p_{s} the saturated water pression (Pa), g the acceleration due to gravity (m/s²), and ψ the total water potential in equilibrium with vapour (m).

The vapour diffusion in a porous medium is obviously slower than in free air. Many authors agree in expressing the coefficient D in eq. (7) as a linear function of the medium porosity (% of total volume occupied by the soil particles):

$$D = \alpha \, D_{\mathrm{o}} \, \varepsilon \tag{9}$$

where D_{o} is the vapour diffusion coefficient in free air (=0.611×10⁻⁴.m²/s, ε the soil porosity, and α a constant value: $0.58 - 0.67$ (e.g., Tschapek *et al.*, 1966).

The density and porosity of disperse bodies, including soil, is an index of the mutual arrangement of the particles, their mutual proximity, and the degree of contact. For ideal systems of balls, independently of the systems of packing (cubic, hexagonal, tetrahedral, n-stage hexagonal, ...), the relationship between the density and porosity is expressed by the following formula (e.g. Verschinin *et al.*, 1966):

---

[1] The wilting point is an extremely important parameter of soil. The store of water up to such a moisture content is absolutely unavailable to plants.

$$\varepsilon = \left(1 - \frac{\rho}{\rho_s}\right) 100 \tag{10}$$

where ρ represents the apparent density (mass of the particles referred to the total volume of the disperse body, including pores), and ρ_s the so called specific gravity of particles. This latter is defined as the product of the true density (mass of the dry particles referred to the volume occupied by the solid phase) by the acceleration of gravity. A common value for the ρ_s of soils is 2700 kg/m³.

According to Vershinin *et al.* (1966), it may be generally stated that the entire variety of pores and densities occurring in natural soils can be explained by a certain ideal packing of soil particles. If the apparent density of soil and the specific gravity of its constituent particles are known, the total porosity of the system can be then estimated by eq. (10).

As mentioned previously, the store of water up to the wilting point is absolutely unavailable to plants. On the other hand, moisture contents above the soil field capacity (soil capillary water capacity) may cause root damage (asphyxia). We will then assume that soil moisture content is always between these two extremes. To estimate the total water potential in equilibrium with vapour, χ, the following linear functional relation will then be used:

$$\psi = \psi_{wp} + \frac{\psi_{fc} - \psi_{wp}}{\omega_{fc} - \omega_{wp}} (\omega - \omega_{wp}) \tag{11}$$

ϖ_{wp} is the volumetric moisture content at the wilting point, where $\chi = \chi_{wp} = -10^{-4.15}$ cm, and ϖ_{fc} is the volumetric moisture content at field capacity, with $\chi = \chi_{fc} = -10^{-2.71}$ cm.

The model describing the soil thermal behaviour is then defined by eqs (4) to (11), with boundary conditions of Dirichlet type:

$$\begin{cases} T(z=0,t) = T_{soil,top}(t) \\ \\ \\ T(z=L,t) = T_{soil,bottom}(t) \end{cases} \tag{12}$$

with $T_{soil,top}$ and $T_{soil,bottom}$ the temperatures imposed at the top and bottom of the soil layer. As in the previous case (support model), assuming Dirichlet boundary conditions, the soil behaviour is decoupled from the other roof components.

Model input parameters are the thickness of the soil layer (L), the soil apparent density (ρ), the soil moisture content (ϖ), the soil wilting point (ϖ_{wp}) and the soil field capacity (ϖ_{fc}).

Canopy Model

The canopy is composed of the leaves and the air within the leaf cover. The main processes contributing to the definition of the canopy thermal state are (Fig A.5):

- solar radiation absorbed by the leaves;
- longwave radiative exchanges (TIR) between the leaves and the sky, the leaves and the ground surface and among the leaves themselves;
- convective heat transfer between the leaves and the canopy air, and between the ground surface and the canopy air;
- evapotranspiration in leaves. This process includes three phenomena: water evaporation inside the leaves (stomatal cavity), vapour diffusion to the leaf surface and convective vapour transport from the leaf surface to the air;
- evaporation/condensation of water vapour in the soil surface and vapour convective transfer between the soil surface and the air;
- convective heat and vapour transfer between the air within the canopy and the free air.

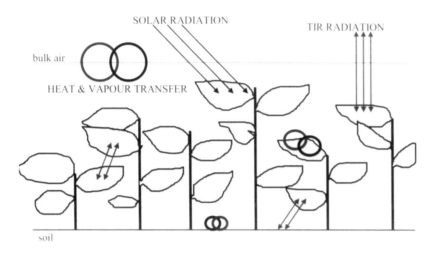

Fig. A.5 Energy transfer processes in a leaf canopy.

The complexity of a canopy as a system of sources and sinks of heat and mass precludes an exact description of its physical behaviour. In aiming for a simpler representation (model) of the canopy, one faces two types of problem. First, the inherent spatial complexity and lack of homogeneity of the foliage. This implies that, for an accurate description, the number of simultaneous equations to be solved could be five times higher than the number of leaves in the canopy. Second, the turbulent nature of the air stream within (and above) a canopy. Its consequence is that the direction and magnitude of the energy and mass fluxes vary at any moment and cannot be exactly predicted.

In this paper the canopy will be regarded as one homogeneous layer, characterized by one value for leaf temperature and one value for the temperature and vapour content of the air within (Fig. A.6). Such a layer is bounded by the ground surface at the bottom and an ideal surface at the top, which is, in turn, homogeneous.

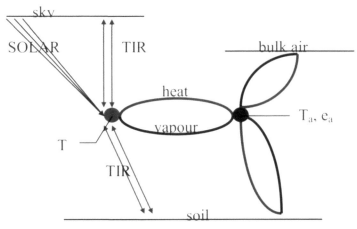

Fig. A6. Single layer representation of a canopy.
SOLAR = Solar radiation flux; TIR = longwave radiation exchanges.

With these assumptions, the macroscopic or "mean" energy and vapour balance equations in a canopy are:

$$
\begin{cases}
(\rho c)_p \, d \, \mathrm{LAI} \, \dfrac{d \, T_p}{d \, t} = \varphi_{rad,sol} + \varphi_{rad,TIR} + \varphi_{conv,p\text{-}a} + \varphi_{trans,p\text{-}a} \\[2ex]
(\rho c)_a \, L \dfrac{d \, T_a}{d \, t} = \varphi_{conv,a\text{-}p} + \varphi_{conv,a\text{-}g} + \varphi_{conv,a\text{-}\infty} \\[2ex]
\rho a \, L \, \dfrac{d \, \theta_a}{d \, t} = \varphi_{vap,a\text{-}p} + \varphi_{vap,a-g} + \varphi_{vap,a\text{-}\infty}
\end{cases}
\tag{13}
$$

where,
T_p is the leaf temperature (average in the control volume) (K);
T_a is the air temperature (average in the control volume) (K);
υa is the air specific humidity (average in the control volume) (kg/kg);
$(\rho c)_p$ is the leaf specific thermal capacity (J/m³/K);
d is average leaf thickness (m);

$(\rho c)_a$ is air specific thermal capacity (J/m³/K);

ρa is the air density (kg/m³);

L is the canopy layer thickness (m);

$\omega_{rad,sol}$ is the solar radiation absorbed by the leaves (W/m²);

$\omega_{rad,TIR}$ is the net thermal radiation flux on leaves (W/m²);

$\omega_{conv,p\text{-}a}$ is the sensible heat flux between the foliage and the canopy air (W/m²);

$\omega_{trans,p\text{-}a}$ is the energy flux due to leaf transpiration (W/m²);

$\omega_{conv,a\text{-}p}$ is the sensible heat flux between the canopy air and the foliage ($\varphi conv,a\text{-}p = -\omega conv,p\text{-}a$) (W/m²);

$\omega_{conv,a\text{-}g}$ is the sensible heat flux between the canopy air and the ground surface (W/m²);

$\omega_{conv,a\text{-}o}$ is the sensible heat flux between the canopy air and the outdoor air (W/m²);

$\omega_{vap,a\text{-}p}$ is the vapour flux between the canopy air and the foliage (kg/m²);

$\omega_{vap,a\text{-}g}$ is the vapour flux between the canopy air and the ground surface (kg/m²);

$\omega_{vap,a\text{-}}$ is the vapour flux between the canopy air and the outdoor air (kg/m²).

The equations below are written per unit of ground area. The total leaf area (one face only) contained in a volume of unit base is named Leaf Area Index (LAI).

Net thermal radiation
The longwave transmittance, $_{\tau l}$(LAI), of a canopy having leaf area index LAI is defined as the quotient between the longwave radiative flux entering either the upper or the lower surface of a section of the canopy and the flux leaving the other end of that section. It is also called "transmittance of a canopy of black leaves" since, in the longwave range, transmittance and reflectance of the leaf tissue are negligible. The canopy will transmit only the radiation which is not even once intercepted by a leaf.

It can be shown (Ross, 1981) that for diffuse longwave radiation, $_{\tau l}$(LAI) is affected only by the geometrical properties of the canopy. Indeed, under some assumptions, the theoretical functions for extinction of radiation in a turbid medium may be applied to a canopy (e.g., Norman, 1975), to yield:

$$\tau_1(\text{LAI}) = \exp\left(-k_1 \text{ LAI}\right) \tag{14}$$

with k_l the extinction coefficient for longwave radiation that can be analytically calculated for some idealised leaf angle distributions. Some values for it, deduced from the literature (see, e.g., Stanghellini, 1983), are supplied in Table A.1.

TABLE A.1. VALUES OF THE EXTINCTION COEFFICIENT FOR LONGWAVE RADIATION

LEAF DISTRIBUTION	k_l
Horizontal	1; 1.05
Conical ($\alpha = 45°$)	0.829
Vertical ($\alpha = 90°$)	0.436
Spherical	0.684; 0.81

The values in Table A.1 are as deduced from the literature, for idealized leaf angle distributions. The angle between the leaves and a horizontal plane is represented by the symbol α.

A canopy absorbs a fraction equal to $(1-\tau_l)$ of the longwave radiation it receives. The net thermal radiation flux in a canopy is then written as:

$$\varphi_{\mathrm{rad,TIR}} = (1-\tau_1)\left[\underbrace{\sigma\, T_{\mathrm{sky}}^4}_{\text{TIR received from the sky}} + \underbrace{\sigma\, T_g^4}_{\text{TIR received from the ground}} - \underbrace{2\,\sigma\, T_p^4}_{\text{TIR emitted by the canopy}}\right] \tag{15}$$

where T_{sky} is the sky temperature and T_g the temperature of the ground surface.

Net solar radiation

The shortwave radiation transmitted by a canopy is the sum of the unintercepted radiation and the radiation that is either transmitted or reflected downwards (or both) by any leaf within the canopy. The shortwave transmittance $\tau_s(\mathrm{LAI})$ for diffuse shortwave radiation of a canopy with a leaf area index LAI, can be represented (Kasanga and Monsi, 1954), with reasonable accuracy, by an exponential law:

$$\tau_s = \exp\left(-k_s\,\mathrm{LAI}\right) \tag{16}$$

The coefficient of extiction ks has to be a function of the optical properties of the leaves, and in Goudriaan (1977) is calculated through:

$$k_s = \left[\left(1-\tau_t\right)^2 - \rho_t^2\right]^{1/2} k_1 \tag{17}$$

with τ_t and ρ_t the transmittance and reflectance of the leaf tissue, respectively. For a "mean" green leaf, this equation results in $k_s = 0.74\, k_1$ (Ross, 1975). For species with mainly horizontal leaves we have $k_s \cong 1.10$, and for species with mainly vertical leaves we get $k_s \cong 0.29$.

The reflectance of a canopy is always smaller than that of the leaves composing it. The mutual shading of leaves and the multiple scattering within the canopy result in a "cavity" effect, which causes an additional absoption of radiation. It will be assumed (Stanghellini, 1983) that the canopy behaves as a dense canopy (stand of unitary interception for diffuse radiation, or completely covering the soil) with respect to the fraction of the incident radiation it intercepts. Hence:

$$\rho_s(\mathrm{LAI}) = \left(1-\tau_1(\mathrm{LAI})\right)\rho_\infty \tag{18}$$

where ρ_∞ is the reflectance of a dense canopy.

A canopy absorbs a fraction equal to $1-\tau_s-\rho_s$ of the shortwave radiation it receives. The net solar radiation flux absorbed by a canopy is then written as:

$$\varphi_{\mathrm{rad,sol}} = \left[1 - \tau_s - \left(1-\tau_s\right)\rho_\infty\right]\left(1+\tau_s\,\rho_g\right)\varphi_s \tag{19}$$

where ω_s represents the solar radiation at the top of the canopy, and $\tau \rho_g \varphi_s$ the solar radiation reflected by the ground.

Convective heat transfer between leaves and air

The transfer of sensible heat between the foliage and the air within the canopy will be represented by the Newton law of convection[2]:

$$\varphi_{conv,p-a} = -\varphi_{conv,a-p} = -2\,LAI\,\frac{\rho c_p}{r_e}\left(T_p - T_a\right) \tag{20}$$

where r_e indicates a mean canopy resistance to sensible heat transfer, or "canopy external resistance" which is, in fact, defined by eq. (20). The experimental data available in the literature are not conclusive about a unique equation to predict external resistance of leaves exposed to low windspeeds, in the presence of external sources of turbulence, and being only a few degrees warmer or cooler than the sourrounding air. We will adopt here the correlation model proposed by Stanghellini (1987), which is based on exhaustive experimental work:

$$r_e = \frac{a\,l^m}{\left(1\left|T_p - T_a\right| + b\,u^2\right)^n} \quad (s/m) \tag{21}$$

where l is the leaves characteristic length and u the wind speed. a, b, m and n are empirical coefficients (a = 1174, b = 207, m = 0.5, n = 0.25 for tomato crops). Using eq. (21), Stanghellini (1987) shows that for increasing wind speed or dimension, the consequence of an error in the estimation of the external resistance reduces. For instance, variations of wind speed above say, 0.2 m/s, have a small effect on the external resistance, whereas the effect will be much greater for air flow in the range of 0.05 – 0.1 m/s.

Transpiration flux

If vapour pressure is chosen as an appropriate forcing function, the energy flux consumed to allow water to evaporate in leaves may be represented by a law analogous to that in the previous section:

$$\varphi_{trans,p-a} = -2\,LAI\,\frac{\rho c_p}{\gamma\left(r_e + r_i\right)}\left(e_p - e_a\right) \tag{22}$$

where γ is the thermodynamic psychrometric constant (Pa/K), e_p and e_a (Pa) are the vapour pressure at the leaf surface and at the canopy air, respectively, r_e is the canopy external resistance (s/m), defined by eq. (20), and r_i is the so called internal resistance to vapour transfer of the canopy, or bulk stomatal resistance, which is, in fact, defined by eq. (22).

---

[2] The heat flux density, ω (W/m²), through a finite surface in contact with a fluid is assumed to be proportional to the difference of the internal energy concentration, $\rho c_p T$ (J/m³), between the surface and the fluid: $\varphi = \kappa \rho c_p (T_{surface} - T_{fluid})$. The proportionality constant, κ (m/s), is called the "bulk heat transfer coefficient", which is related to the heat transfer resistance, r (s/m), by: $\kappa = 1/r$. The commonly used convective coefficients, h (W/m²/K), which are also called "bulk heat transmission coefficients", are then defined by: $h = \kappa \rho c_p = \rho c_p / r$.

The internal resistance of a canopy is known to be affected by a number of physiological and environmental parameters. Only the latter group will be considered here. Among the environmental parameters, shortwave irradiance (ω_s) appears to be the most important. Leaf to air vapour pressure difference ($e_p - e_a$), leaf surface temperature (T_p) and CO_2 concentration of the air are known to also play a significant role.

A very interesting discussion about the physical meaning of the internal resistance as well as about its functional relations with each driving variable, can be found in Stanghellini (1987), Cowan and Troughton (1971) and Farquhar and Sharkey (1981). Stanghellini (1987) shows that the apparent behaviour of the internal resistance of a canopy is almost exclusively determined by the internal resistance of the leaves composing it. Accordingly, the observed behaviour of the internal canopy resistance can be represented by a phenomenological model quite similar to the one suggested for the internal resistance of a leaf:

$$r_i = r_{min} \, \tilde{r}_i(\varphi_s) \, \tilde{r}_i(T_p) \, \tilde{r}_i(CO_2) \, \tilde{r}_i(e_p - e_a) \tag{23}$$

where r_{min} (s/m) is the minimum possible resistance, whose magnitude has a purely physiological origin. The symbols r_i represent dimensionless functions larger than unity, quantifying the relative increase of the internal resistance, whenever one of the parameters is limiting the rate of transfer of water vapour. The following function was proposed to represent such effects:

$$\begin{cases} \tilde{r}_i(\varphi_s) = \dfrac{\overline{\varphi}_s + C_1}{\overline{\varphi}_s + C_2}; \; C_1 > C_2; \; \overline{\varphi}_s = \dfrac{\varphi_s}{2 \, LAI} \\[2mm] \tilde{r}_i(T_p) = 1 + C_3 \left(T_p - T_m\right)^2 \\[2mm] \tilde{r}_i(CO_2) = 1 + C_4 \left(CO_2 - 200\right)^2 \\[2mm] \tilde{r}_i(e_p - e_a) = 1 + C_5 \left(e_p - e_a\right)^2 \end{cases} \tag{24}$$

For a tomato crop, the following values are proposed in Stanghellini (1987): r_{min} = 82 (s/m), C_1 = 4.3, C_2 = 0.54, T_m = 24.5 (°C), C_3 = 2.3e–2, C_4 = 6.1e–7 and C_5 = 4.3. For sensitivity analysis purposes, the internal resistance of a canopy will be represented by:

$$r_i = f \, r_{i,tomato} \tag{25}$$

the coefficient f allowing us to represent a canopy evaporating more ($f<1$) or less ($f>1$) than a tomato crop.

Heat and vapour flux between the ground surface and the air

The transfer of sensible heat between the ground surface and the air within the canopy will be represented by the Newton law of convection:

$$\varphi_{conv,g\text{-}a} = -\varphi_{conv,a\text{-}g} = -h_g \left(T_g - T_a\right) \tag{26}$$

where T_a is the temperature of the air canopy, T_g the temperature of the soil surface, and h_g (W/m²/K) the bulk convective coefficient of heat transport.

In a similar way, the transfer of water vapour between the ground surface and the air within the canopy will be represented by:

$$\varphi_{vap.g\text{-}a} = -\varphi_{vap.a\text{-}g} = -\tilde{h}_g (e_g - e_a) \qquad (27)$$

where e_a is the vapour pressure of the canopy air (Pa), e_g the vapour pressure at the soil surface (Pa) and h_g (kg/s/m²/Pa) the bulk convective coefficient of vapour transport.

Since heat as well as vapour transport is convective and not diffusive (Lewis number close to 1), we can assume

$$\tilde{h}_g = \frac{1}{\Lambda\gamma} h_g \qquad (28)$$

where λ is the thermodynamic psychrometric constant (Pa/K) and Λ the latent heat of vaporization (J/kg).

Heat and vapour transport between the air within the canopy and outdoor air

Heat and vapour fluxes caused by mass transfer between the canopy air and the air outside the canopy boundary layer, will be described by

$$\varphi_{conv,a\text{-}\infty} = -h_{a\infty} (T_a - T_\infty) \ \ (\text{W/m}^2); \qquad h_{a\infty} = R\,L\,\rho c_p \ \ (\text{W/m}^2/\text{K}) \qquad (29)$$

and

$$\varphi_{vap,a\text{-}\infty} = -\tilde{h}_{a\infty} (e_a - e_\infty) \ \ (\text{kg/m/s}); \qquad \tilde{h}_{a\infty} = \frac{1}{\Lambda\gamma} h_{a\infty} \ \ (\text{kg/m}^2/\text{s/Pa}) \qquad (30)$$

respectively, with T_∞ the outdoor air temperature and e_∞ the outdoor air vapour pressure. L is the canopy height and R represents the air exchange rate (per s). This approach is quite similar to that used for describing heat flux by ventilation in buildings.

The canopy model is then defined by eq. (13) to (30). The model inputs are solar radiation flux (ω_s), sky temperature (T_{sky}), ground surface temperature (T_g), dry-bulb outdoor temperature (T_∞) and outdoor air vapour pressure (e_∞). The total number of parameters in the model is 21.

Green Roof Model

The green roof model is composed of the support model, the soil model, and the canopy model, together with the coupling models, which will be described here. Such models represent the real boundary conditions at the canopy–soil and canopy–support interfaces, satisfying the physical constraint of continuity for the states' variables and the flux densities.

Canopy–soil coupling

Continuity of the state variables at the canopy–soil interface, implies

$$T_g(t) = T_{\text{soil,top}}(t)$$

$$(31)$$

$$e_g(t) = p_v(T_{\text{soil,top}}, \omega_{\text{soil,top}})$$

with the partial water pressure p_v calculated by eq. (8) at the temperature and moisture content of the soil top surface.

Continuity of the heat flux density at the canopy–soil interface, implies

$$-\left(\lambda + \Lambda D_{vT}\right)\left.\frac{\partial T(z,t)}{\partial z}\right|_{z=0} = \underbrace{(1-\rho_g)\,\tau_s\,\varphi_s}_{\text{absorbed solar radiation}}$$

$$+ \underbrace{\left[\tau_l\,\sigma T_{\text{sky}}^4 + (1-\tau_l)\,\sigma T_p^4 - \sigma T_g^4\right]}_{\text{net thermal radiation flux}} \quad (32)$$

$$- \underbrace{h_g\,(T_g - T_a)}_{\text{sensible heat flux}} - \underbrace{\Lambda\,\tilde{h}_g\,(e_g - e_a)}_{\text{latent heat flux}}$$

This equation represents the energy balance at the soil top surface, when no energy storage is assumed.

Continuity of the vapour flux density at the canopy–soil interface, implies

$$-D_{vT}\left.\frac{\partial T(z,t)}{\partial z}\right|_{z=0} = -\tilde{h}_g\,(e_g - e_a)$$

$$(33)$$

Soil–support coupling model

In a similar way, for the soil–support interface we can write

$$T_{\text{soil,bottom}}(t) = T_{\text{support,top}}(t) \qquad \text{Continuity of temperature}$$

$$(34)$$

$$-\left(\lambda + \Lambda D_{vT}\right)\left.\frac{\partial T(z,t)}{\partial z}\right|_{z=L} = -\lambda_s\left.\frac{\partial T_s(z,t)}{\partial z}\right|_{z=0} \qquad \text{Continuity of heat flux}$$

The roof support is supposed to be waterproof and impermeable to water vapour.

The green roof model has been translated under an "object-oriented" form, and implemented in MATLAB. It is composed by three modules (canopy, soil and support models) limited by Dirichlet boundary conditions on their frontiers, where temperature (or moisture content) is fixed, and two interfaces (canopy–soil and soil–support models) where the relations between two connected modules are described. Each module has frontiers to communicate with the outside. A frontier is a set of some input–output variables (e.g., temperature and heat flux at the top surface of the support model). A detailed description of the adopted modelling approach can be found in Ebert *et al.* (1991) and Ebert (1993). The solution method is based on the assumption that a global solution of the problem may be achieved by a set of local resolutions and relations traducing the constraints expressed on the coupling interfaces Ebert (1993). Each interface sends constraints to the modules to be coupled, which attempt to reach a state satisfying these constraints. In particular, a Newtonian iteration algorithm has been used to find the interface state variables that satisfy the contraints of flux densities continuity of the connected modules. Differential equations inside the modules (domaine state variables updating) are integrated using a method belonging to the general family of BDF (Backward Differentiation Formulas) methods, specially adapted for solving difficult ordinary differential equations (Shampine). The finite volumes method has been used to translate continuous models (soil and support models) into space discrete models.

References

Andrews, R.V. (1955) Solving conductive heat transfer problems with electrical analogue shape factors. *Chemical Engineering Progress*, 51(2): 67.

Cowan, I.R. and Troughton, J.H. (1971) The relative role of stomata in transpiration and assimilation. *Planta '97*.

De Vries, D.A. (1958) Simultaneous transfer of heat and moisture in porous media. *Transactions of the American Geophysical Union*, 39.

Ebert, R., (1993) Développement d'un environnement de simulation de systèmes complexes. Application au bâtiments. Ph.D., Ecole Nationale des Ponts et Chaussées, Paris.

Ebert, R., Peuportier, B. and Lefebvre, G. (1991) Simulation of the thermal building behavior based on an object-oriented ADA implementation. *Conference Proceeding Building Simulation '91*. IBPSA, Sophia-Antipolis, France.

Farquhar, G.D. and Sharkey, T.D. (1981) Stomata and photosynthesis. *Annual Review of Plant Physiology* 32.

Goudriaan, J. (1977) *Crop Micrometeorology: A Simulation Study*. Simulation Monographs, Pudoc, Wageningen.

Hinton, E. and Owen, D.R.J. (1979) *An Introduction to Finite Element Computations*, Pineridge Press.

Kasanga, H. and Monsi, M. (1954) On the light transmission of leaves, and its measuring for the production of matter in plant communities *Japanese Journal of Botany*, 14.

Luikov, A.V. (1966) *Heat and Mass Transfer in Capillary Porous Bodies*. Pergamon Press, Oxford.

MATLAB, High-performance numeric computation and visualization software. The MathWorks, Inc.

Molina, J.L. and Rodriguez, E.A. (1998) Detailed modelling of water ponds. In *Final Research Report of Task 2: Modelling Cooling Roofs and Design Tool Development*. ROOFSOL Project European Commission Joule Programme JOR3CT960074.

Norman, J.M. (1975) Radiative transfer in vegetation. In de Vries, D.A. and Afgan, N.H. (Eds) *Heat and Mass Transfer in the Biosphere*, Scripta Bokks Company, Washington.

Palomo Del Barrio, E. (1998a). Analysis of the green roofs cooling potential in buildings. In *Final Report Task 2 ROOFSOL: Roof Solutions for Natural Cooling*, Contract no: JOR3CT960074, Commission of the European Communities, DG XII Science, Research and Development.

Palomo Del Barrio, E. (1998b). Analysis of the green roof cooling potential in buildings. *Energy and Buildings* 27: 179–193,

Philip, J. and de Vries, D.A. (1957) Moisture movement in porous materials under temperature gradients. *Transactions of the American Geophysical Union*, 38.

Ross, J. (1975) Radiative transfer in plants communities. In Monteith, J.L. (Ed.) *Vegetation and Atmosphere*. Academic Press, London.

Ross, J. (1981) *The Radiation Regime and Architecture of Plant Stands*, Dr. Junk Publishers, The Hague.

Shampine, L.F. and Reichelt, M.W. The Matlab Internet services. www.mathworks.com.

Stanghellini, C. (1983) *Radiation Absorbed by a Tomato Crop in a Greenhouse*. Institute of Agricultural Engineering (IMAG), Wageningen, Research Report 83–5.

Stanghellini, C. (1987) Transpiration of greenhouse crops. Ph.D. Dissertation, Agricultural University, Wageningen.

Thorton, D.E. (1977) *Calculation of Heat Loss from Pipes in Utilities Delivery in Arctic Regions*. EPS, Environment Canada, EP53–WP–77–1: 131–150.

Tschapek, M.W. (1966) *El agua en el suelo*, C.S.I.C., Manuales de ciencia actual 2.

Vershinin, P.V. *et al*. (1966) Fundamentals of agrophysics. In Ioffe, A.F. and Revut, I.B. (Eds) *Israel Program for Scientific Translations*.

Whitaker, S. (1977) Simultaneous heat, mass and momentum transfer in porous media. A theory of drying in porous media. *Advances in Heat Transfer* 13.

Zienkiewicz, O.C. (1980) *El Método de los elementos finitos*. Reverté.

DESIGN SUPPORT SOFTWARE

INTRODUCTION

This appendix describes the RSPT software that was developed for simulating the performance of roof cooling techniques, which is included on a CD with this publication.

Given appropriate inputs by the user the software performs hourly thermal simulations for the selected building configuration and roof cooling technique, as well as for a reference building for comparison that is generated automatically. The output is in the form of predicted energy savings compared with an air-conditioned reference building, or as a measure of improved thermal comfort for free-running buildings. The simulations can be performed for any of the locations for which a weather file is provided in the RSPT database. Alternatively, the output can be in the form of regional maps, as was illustrated in previous chapters. These provide comparative information on the relative performance of roof cooling techniques over a wide geographic area. At present the maps are drawn mainly for parts of Europe, but they could be extended to other regions.

The main sections of this user guide are:

- computer requirements and installation of the software;
- main characteristics of the RSPT simulation code;
- how to use the software;
- importing weather files.

SYSTEM REQUIREMENTS AND INSTALLATION

The RSPT software was developed for Windows operating systems. The recommended computer specification is:

- processor: Intel Pentium IV 2.0 GHz or equivalent;
- main memory: 256 MB;
- video adapter: 32 MB AGP;
- CD ROM driver.

The software may also run on systems with a lower specification than above, but calculations will take longer. To install the software insert the CD-ROM in your CD drive and run the setup program iRSPT.exe. The installation process will place the software files in a folder named RSPT. Users of the Meteonorm Global Meteorological Database (Meteotest, 2003; www.meteonorm.com) should copy the Meteonorm2Met.exe utility from the CD to the RSPT root folder. This utility will allow them to convert weather files generated with Meteonorm to the format required for use with RSPT.

THE RSPT SIMULATION CODE

Building Modelling

The main characteristics of the building modelling methodology used by the RSPT software are:

- Heat transfer through opaque elements is calculated by means of response factors, which are in turn calculated by a detailed finite difference model for transient heat transfer. Heat transfer is considered as one-dimensional in walls and slabs, and two-dimensional in elements below ground.

- Heat transfer through transparent and semi-transparent elements is calculated in steady state, using the solar and thermal transmittance values given by product manufacturers.

- Calculation of solar gains through the building envelope takes account of shadows cast by external obstructions and the building's own surfaces.

- Heat transfer involving roof ponds, radiators and cooling panels is based on the models presented in Appendix A; planted roofs are not treated by the RSPT software.

The program's energy balance equations calculate surface and air temperatures as a function of hourly meteorological parameters and inputs by the user. The calculation of the indoor air temperature takes account of internal heat gains, energy exchanges by ventilation and air infiltration, and the output of mechanical heating, ventilation and air-conditioning (HVAC) systems. The RSPT software solves the balance equations iteratively at each time step in the following sequence:

- the effect of an HVAC system is simulated under idealised conditions or as described by a proportional control law with inputs of surface and air temperatures as calculated in a previous time step;

- surface temperatures are calculated successively for the following time step with all other parameters known;

- the indoor air temperature is calculated with all other parameters known as above.

In this approach two to four iterations are required at each time step. Because of the nature of the calculation techniques used, time steps need to be short in order to obtain accurate results. Typically, they are set at 10-minute intervals whilst meteorological magnitudes are assumed constant over hourly intervals, as read from the meteorological file for the location.

Comparative Performance

Given the description of a user-defined building and weather data in appropriate format an RSPT run provides estimates of energy savings and thermal comfort improvements from roof cooling techniques by performing two sequential simulations. When the user selects to assess energy savings the software simulates the user-defined building with the roof cooling application selected plus an HVAC system. Next it simulates the same building description with the same HVAC system but with a user-defined reference roof in place of the roof cooling system. The reference roof represents the roof construction that would have been specified normally instead of a roof cooling application. The difference in the energy consumption predicted for these two cases represents the energy saving. When the user selects assessment based on occupant thermal comfort, the simulations are performed without an HVAC system. In this case the building's indoor air temperature is calculated with the roof cooling application in place. The calculation is then repeated for an identical building, but with a user-defined reference roof. The improvement, if any, in occupant thermal comfort is measured as a reduction in the number of hours by which the simulated indoor air temperature exceeds a set of predetermined set temperatures selected to represent common upper limits of thermal comfort: 23°C, 25°C and 27°C. The reduction in the mean and peak indoor temperatures is also provided in the summary output.

The RSPT tool is intended to be used in the following two ways:

1. to investigate the possible energy savings or comfort improvements from alternative roof cooling systems on user-defined building configurations for a given location;

2. to provide comparative applicability indicators by generating regional performance maps for different techniques over a wide geographical area.

Regional performance maps are produced by the software by the following sequence of operations:

1. *Calculations for base locations:* paired simulations (of buildings with and without a roof cooling application) are performed for a number of base locations representing the range of different climatic conditions in the given region (currently the software generates the maps for parts of Europe). Hourly data representing a typical meteorological year are used for the base locations.

2. *Generation of correlations:* the simulation results for the base locations are correlated with monthly mean climatic data of the base locations.

3. *Extrapolation:* the correlation algorithms obtained are used to calculate performance parameters for additional locations using mean climatic data as correlation parameter. The database currently held in the software includes such data for over 100 European locations, evenly spread across the continent.

USING THE SOFTWARE

Building Definition

The building is defined in plan on a sketchpad in the opening screen of the software (Fig. B.1). The coordinate system is oriented with positive values of the Y-axis pointing to the north and positive values on X-axis to the east. Orientation can be modified in later steps if required (see **General Data** below). All building dimensions are assumed to be in metres. The scale of the sketchpad may be changed by entering a different value for '**Cell size**', thus changing the distance between adjacent points on the grid. The x, y coordinates of the mouse cursor are displayed continuously on the screen.

A building may consist of one or more storeys that are displayed as polygons on the sketchpad. Each storey can be given a different floor-to-ceiling height if required. Each side of the polygon represents an external wall. New wall vertices may be added, and existing ones relocated, by entering their respective coordinates in the vertex table at the left of the sketchpad.

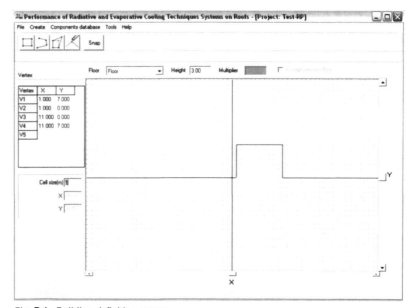

Fig. B.1 Building definition screen.

External walls without windows are represented by thin red lines. After the properties of the wall have been entered and confirmed by the user, the graphic representation changes to a thick red line. External walls with one or more windows are represented by a thick yellow line. In the example shown (Fig. B.1), the walls with north, east and south orientations have one or more windows, whereas the one oriented to the west has no windows. Elements selected by the user change colour, from red or yellow to green, for the duration of their selection.

The graphic editing tool consists of five buttons:

Create polygon: with this button selected a rectangle can be drawn by clicking the left mouse button at the desired location for one vertex, then dragging the mouse with the button still depressed to the opposite corner. (Note: If this button is selected when a polygon is already present on the sketchpad, the existing polygon will be replaced by the new rectangle.)

Add vertex: click on the sketchpad to indicate the location of the desired vertex. When a vertex is added, the nearest existing wall is split into two. Additional vertices may be added either inside or outside the existing polygon.

Move vertex: indicate the desired location of a specific vertex on the sketchpad with the cursor. The nearest existing vertex will be relocated to the specified position. If a vertex is to be moved to a position that is not adjacent to its existing one, it may be dragged to the new location by depressing the left mouse button until the cursor is in the desired location.

Delete vertex: with this button, unwanted vertices may be deleted by left-clicking the mouse near their position.

Snap button: initiates snap mode so that polygon vertices can be created only on the dots of the workspace grid.

General Data

Right-clicking the mouse in the sketchpad area outside the building plan opens the **General Data** menu (Fig. B.2). This menu allows a name to be entered for a simulation run. The name of the location is chosen from the list held in the location database (locations for which weather files are available); see later in this Guide on how to extend to location database and create weather files for different locations. A reference azimuth must be entered in the following box if the orientations of the building elevations are different from those identified on the graphic interface. The building azimuth is defined as the angle between the geographical north and the positive direction of *Y* axis on the graphic interface, where the clockwise direction is considered positive. The starting date and duration of the calculation period must be entered in the remaining boxes. In early design stages select a period in mid-summer of 10–15 days; this should be adequate to provide acceptable results in the minimum computational time. Running the software for periods shorter than 10 days may not provide accurate results.

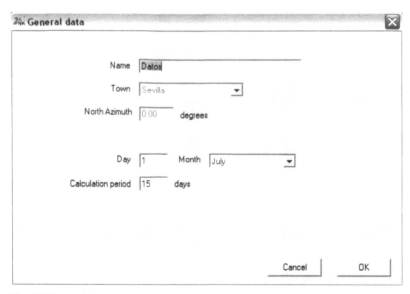

Fig. B.2 General data menu.

For best accuracy the software may be run for a complete summer season (some 120 days for European locations); this may entail very long calculation times.

Operational Conditions

A right-click of the mouse in the sketchpad outside the building plan also allows selection of the **Operational Conditions** menu (Fig. B.3). Here the user selects either *Residential* or *Office* use for the software to load appropriate occupancy schedules and internal heat gain profiles. Another menu provides choice between *Conditioned* (indoor conditions controlled by HVAC system) or *Unconditioned* (free-running building). A set of default values is applied for the conditioned category.

Building Elements

Right-clicking near an element on the sketchpad screen prompts a pop-up menu with options for the specification of external building elements. Selecting the *Envelope* definition option opens a new window from which different specifications of building elements can be chosen, and openings can be added to an external wall if required (Fig. B.4). There are two classes of walls: *Standard* or *Adiabatic*. The Standard option must be selected for external walls. The Adiabatic option should be selected for internal walls. Each option leads to a drop-down list for selecting a constructional specification and entering dimensions of the elements. Openings such as a window or door may be added to an opaque wall from the appropriate drop-down list offered by the program. The size and position of openings

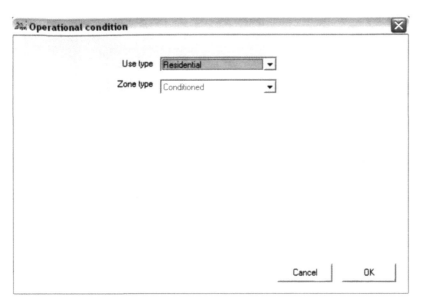

Fig. B.3 Operational conditions menu.

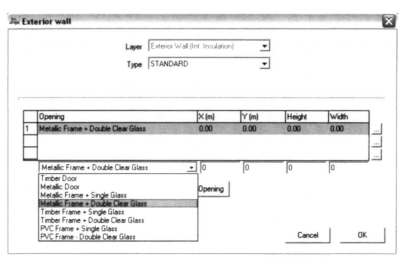

Fig. B.4 Building elements menu.

are specified by entering values for *X*, the horizontal distance from the window to the wall and for *Y*, the sill height.

Right-clicking with the cursor inside the building plan, but away from the wall lines, allows access to menus for the specification of floors and ceilings. There are two options for floors. A floor slab may be either on the ground or below grade (Fig. B.5). For floors that are below grade, the depth of the basement can be typed in the appropriate box.

Roofs must be defined twice. For the reference building the roof should be defined in the same way as an external wall and selecting one of the specifications listed in the menu. The Roof and Floor menus are also accessible by right-clicking inside the building plan. For the test building, a roof cooling system is selected from the relevant menu (**Components Data base**); see following sections.

Additional floors may be added to the building model by selecting the option **Floor, New floor** from the **Create** tab in the main menu. Floors added this way appear listed in the Floor menu on the main screen. Each additional floor has, by default, the same dimensions as the preceding one. However, wall attributes and openings must be entered separately for each floor. To create a number of identical floors, type an appropriate value in the **Multiplier** box above the sketchpad.

When the building being modelled has two or more floors, a cooling panel may be specified in any of the intermediate floor slabs; see **Cooling panels** below.

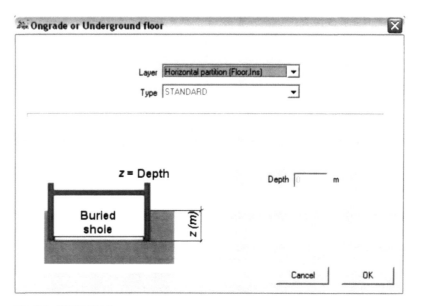

Fig. B.5 Floors menu.

Cooling Panels

The properties of cooling panels are specified on a menu accessed by a right-click inside the building plan area (Fig. B.6). Default values are provided by the software. These can be modified as follows:

- A cooling panel may be either standard or adiabatic. When specified as standard, thermal exchanges will be assumed to take place normally between the two spaces separated by the panel. When the panel is specified as adiabatic, it is assumed that there will be no thermal exchanges between the two spaces, regardless of their temperatures.
- The horizontal distance between the water pipes in the panel can be specified at 10 cm, 20 cm or 30 cm. All variants have thermal insulation above the cooling pipes.
- A water inlet temperature may be specified if the cooling panel is supplied with water at a constant temperature from an external source (such as a river); when the cooling panel is specified in conjunction with a roof pond or a water radiator, the water inlet temperature cannot be specified by the user, but is assumed by the software to be equal to the pond temperature, or to the outlet temperature of the water radiator if connected to the cooling panel directly.
- The total water flow through the cooling panel must be specified: if the cooling panel is coupled with water radiators this must match the water flow through the radiators.
- The diameter of the water pipes in the cooling panel, their length and the number of pipes must be specified; RSPT does not check the dimensional accuracy of these parameters.
- The operating schedule of the cooling panel can be selected from the options provided: for simplicity the duration of 'Daytime' and 'Night' options is kept at a constant 12 hours by the software.
- The box "Next to space" displays a name for the space above the cooling panel.
- The options Roof Pond and Water Radiator are activated when an appropriate component has been selected from the Components database menu.

Roof Cooling Systems

Air Radiators, *Water Radiators* or *Roof Ponds* can be selected and specified from the *Components Database* menu (Fig. B.7). Each of the respective submenus allows the user to define new components (*New*), edit previously saved data (*Change*) or delete a component from the database (*Delete*). When the building has more than one floor, it is also possible to select one or more *Cooling Panels* to operate in conjunction with the

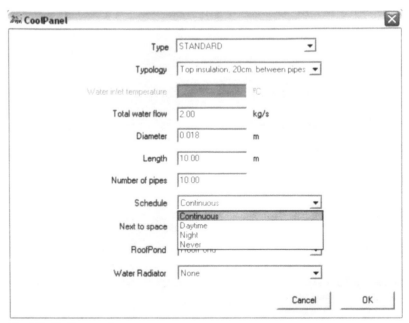

Fig. B.6   Cooling panel menu.

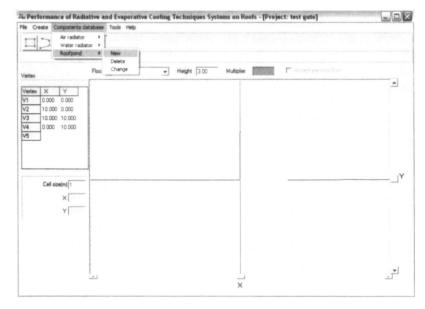

Fig. B.7   Components menu.

selected cooling technique. Water radiators cannot be selected directly. The water radiator menu allows entering and editing a specification once this component has been activated from within a Roof Pond or a Cooling Panel menu.

Air radiators

The properties of the radiating surface including thickness, thermal conductivity and longwave emissivity are fixed and cannot be modified. The following data are open to user inputs (Fig. B.8):

- air flow rate of the radiator in kg/s per metre width of radiator;
- dimensions of the radiator: length (in the direction of flow) and width;
- channel thickness;
- schedule of air circulation.

Water radiators

The diameter of the pipes is fixed (10 mm internal diameter). The schedule for circulating water through the radiators is assumed to take place during night-time only, at a constant rate. The following can be specified by the user (Fig. B.9):

- the initial water inlet temperature; after the first iteration this is assumed to be equal to the temperature of the cooling panel or roof pond to which the water radiator is connected;
- the water flow rate is input in kg/s per pipe; a water flow rate of 0.03 kg/s per pipe results in a water velocity of about 0.4 m/s;
- the number of pipes per metre of radiator width is limited to a maximum value of 50;
- the distance between pipes is calculated automatically once the number of pipes per metre width of radiator has been entered and cannot be modified;
- the overall radiator length and radiator width must be input; values entered here do not necessarily have to correspond to actual building dimensions.

Roof Ponds

The following can be specified by the user (Fig. B.10):

- the depth of water in the pond and the parameters of the spraying system, including jet height, drop radius and number of water changes per hour must be defined;
- the schedule of operation determines whether the spraying system will be operated continuously, during daytime or night only, or switched off;
- the roof pond cover to be in place continuously or during daytime or at night only, or not to be available;

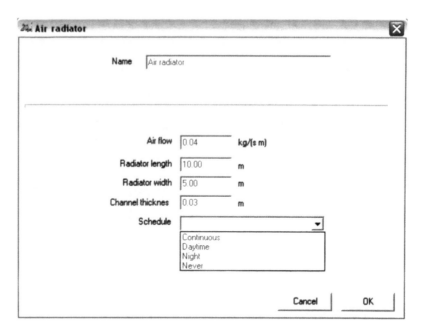

Fig. B.8   Air radiator specification.

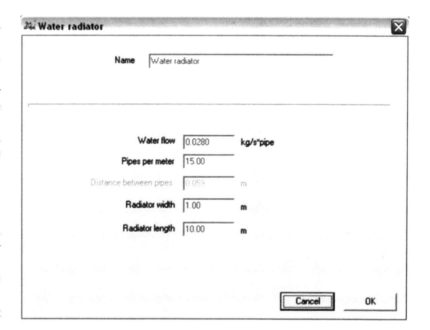

Fig. B.9   Water radiator specifications.

- cover thickness and cover thermal conductivity have default values of 100 mm and 0.036 W/mK, respectively, but these values can be modified prior to saving the roof pond specification;

- selecting to couple the roof pond to a water radiator provides an additional cooling mechanism; if the insulating cover is kept in place continuously and the spraying system is kept inoperative, cooling is assumed to occur by means of the water radiator only. The design parameters and schedule of operation of the water radiators are defined by accessing the relevant menu from the **Component Database** in the main menu.

The schedules for the spraying system and insulating cover are independent of each other.

Calculation and Results

To perform a simulation, the option **Solve** must be selected from the **Tools** menu. This opens the window shown in Fig. B.11. Ticking the box **One location** in the Tools menu will perform the simulations for the location that was specified by the user in the General Data menu. Next, depending on whether the building has been specified with HVAC system or as free-running the **Energy saving** or **Comfort** button should be pressed for the runs. The software will perform paired simulations of the specified building with and without the selected roof cooling technique. For a calculation period of 15 days (see General Data menu above) these will take a few seconds or a few minutes to complete depending on computer power.

The energy savings output gives the respective total energy demands, as well as percentage savings (Fig. B.12). If the system included an evaporative cooling element, the results also include an estimate of water consumption. The thermal comfort results are in the form of percentage occupied hours above the three set points, and the decrease in maximum and average temperatures (Fig. B.13).

When the **One location** box is left unticked, the software will initiate the process of map creation described previously. The calculations for creating a map may take some 30 minutes, depending on computer speed. When the calculations have been completed a new option (**Display**) will appear in the Tools menu. Selecting this option will diplay the map and provide the user with the following possibilities for viewing and saving maps:

Map type: maps may be displayed in either of two colour-coded graphic formats: 2D area fill or contour lines; these are illustrated in Figs B.14–B.17.

Region: the display may be configured to show results for individual countries or regions by making the appropriate selection from the drop-down menu labelled **Region**. This option is the only means of panning across the display, Fig. B.15.

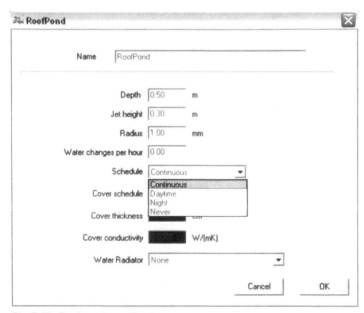

Fig. B.10   Roof pond specification.

Fig. B.11   Initiating calculations.

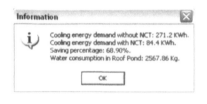

Fig. B.12   Energy output.

Fig. B.13   Thermal comfort output.

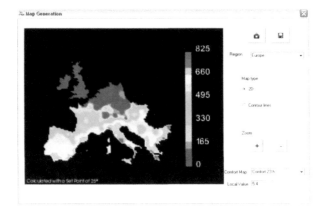

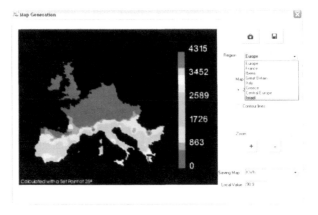

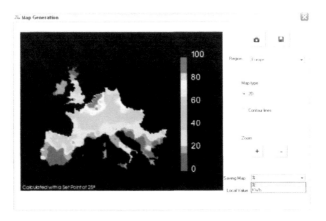

Figs. B.14–B.16 Maps of a section of Europe with colour code identifying areas of varying local potential in energy saving or thermal comfort.

Zoom: the user may zoom in on the current map view using the '+' button or zoom out again using the '-' button. Zooming in or out is carried out relative to the centre of the map. Changing the viewing scale does not change the resolution of the underlying database.

Local value: once a map is obtained, clicking the left button of the mouse at any point on the map allows the user to see the corresponding value in a yellow box in the lower right corner of the screen (Fig. B.15).

In the map showing **Energy savings**, the display may be configured to show either the absolute saving (in units of kilowatt-hours) or as percentage of energy saved through the use of the roof cooling technique specified (Fig. B.16).

When the calculation performed is for **Comfort**, the maps that can be displayed show the number of hours and the equivalent percentage of 24-hour period by which the simulated indoor air temperature exceeded 23°C, 25°C and 27°C, respectively, compared with the reference case.

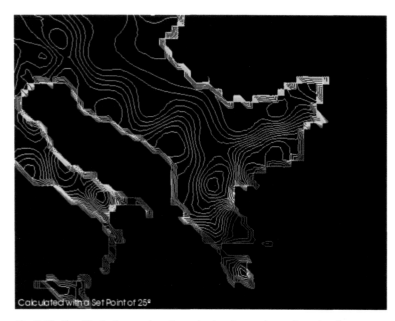

Fig. B.17 Map illustrating contour line output and zoom to a selected region.

Once a map has been obtained, it may be stored for future use, avoiding repetition of the lengthy calculation process. There are two options for doing this (Fig. B16). A bitmap of the graphic display area is generated when the **camera icon** at the upper right corner of the map form is clicked. The resulting file is stored in BMP format and may be edited using image-editing software. An RSPT map file is created when the **floppy disc icon** at the upper right corner of the map form is clicked. The RSPT file can be accessed only from within the RSPT software. The saved map file can be retrieved later by using the option *Load Map* from the *Tools* menu.

In either case, the user can select a name and location for saving the file.

How to Extend the Location Database

Meteotest (2003).
*Meteonorm. Global
meteorological database.*
www.meteonorm.com

The location database of the RSPT software currently contains hourly weather data for a number of European locations including Athens, Bilbao, Catania, Faro, Lisbon, London, Paris and Porto. The full list of weather files can be located in the **Meteorology** folder of the software. Weather data files can be added by the user to this folder. The files should be in ASCII format and the data entries should be provided in the form described below.

The first line should list the name of the location. The second line should have four numbers separated by commas giving the geographic latitude, longitude, altitude and the reference longitude for calculating local time (e.g., $15°E$ for central Europe). These entries should be followed by 8760 hourly records, each of which should list the following data:

- Month number (1–12)
- Day number (1–28/30/31)
- Hour (1–24)
- Dry-bulb temperature (°C)
- Sky temperature (°C)
- Direct solar radiation on horizontal surface (W/m^2)
- Diffuse solar radiation on horizontal surface (W/m^2)
- Absolute humidity (kg/kg dry air)
- Relative humidity (%)
- Wind velocity (m/s)
- Wind direction (degrees: 0 for south, east negative)
- Solar azimuth (degrees: 0 for south, east negative)
- Solar altitude (degrees: 0 to 90)

The solar position must agree with the radiation values. Normally the solar position is given for 30 minutes before the nominal hour. Thus the record for 1200 should give the average radiation incident between 1100 and 1200 hours, and the solar position corresponds to 1130 hours.

Importing Weather Files

RSPT users who have access to the Meteonorm software (Meteotest 2003), may import weather data generated with Meteonorm into RSPT by running the **Meteonorm2Met** utility that can be found on the CD that comes with this publication. This should be copied into the RSPT root folder. The generation and conversion of the weather data follows the two steps described below.

Step 1. Generation of hourly weather data using Meteonorm (v4.10)

Start Meteonorm. In the Format menu select Output Formats and tick the User defined option. In the User defined screen select the output variables to match exactly the form reproduced in Fig. B.18, including ticking the Header option and also Tab under the Hyphen section. Next, open the Units menu in Meteonorm and check that the units for the selected variables are those shown in Fig. B.19. Running Meteonorm's hourly module will then generate a weather data file for the selected location.

Step 2. Conversion of weather data format and calculation of sky temperature using **Meteonorm2Met**

Once the files created with Meteonorm have been saved, the Meteonorm2Met utility can be run (Fig. B.20 and B.21).

- Select the input files (created with Meteonorm).
- Select the destination directory; this should be the Meteorologia folder in the RSPT folder structure.
- Click the button labelled "Start Converting Selected Files". Once the process is completed the green label indicates that the program can be closed.

These operations convert the weather data imported from Meteonorm to the format needed for use with RSPT and will place the converted data file in the Meteorology folder of RSPT. The utility will also calculate hourly values of sky temperature from the parameters given in the Meteonorm output (Meteonorm does not provide the sky temperature).

User defined Output

Time
Year
Hour of the year
Local time
Day of the year

Radiation
Beam
Global radiation, tilted plane
Diffuse radiation, tilted plane
Longwave radiation incoming
Longwave radiation, vertical plane
Longwave radiation outgoing
Global radiation refl.
Radiation balance
Extraterr. radiation
Global luminance
Diffuse luminance
Global radiation, tracked
UVA global
Global radiation horizontal

Temp. / Humidity
Dewpoint temperature
Wet bulb temperature
Surface temperature
Mixing ratio
Driving rain [mm]
Precipitation

Wind / Air pressure
Air pressure

Units (User defined)

☑ Header

Hyphen
○ :   ○ .   ● Tab
○ ;   ○ "   ○ '
Other hyphens [ ]

Output variable
Month
Day of the month
Hour
Air temperature
Direct radiation, horiz.
Diffuse radiation horizontal
Cloud cover fraction
Relative humidity
Wind speed
Wind direction
Solar azimuth
Height of sun

Cancel   OK

Fig. B.18   Meteonorm screen showing output variables that should be selected (right-hand column) for generating weather file to run with RSPT.

Units

Radiation

Monthly val.
● [W / m2]
○ [MJ / m2]
○ [kJ / m2 h]
○ [btu / ft2 h]
○ [kWh / m2]
○ [kWh / m2 d]

Hourly values
● [W / m2]
○ [MJ / m2]
○ [kJ / m2 h]
○ [btu / ft2 h]

Wind
● [m / s]
○ [1/10 m / s]
○ [kt]
○ [km / h]

Air pressure
● [hPa]
○ [1/100 inch Hg]

Temperature
● [°C]
○ [1/10 °C]
○ [F]

Cancel   OK

Fig. B.19   This Meteonorm screen should be checked to ensure that the same units are selected as those shown in the figure

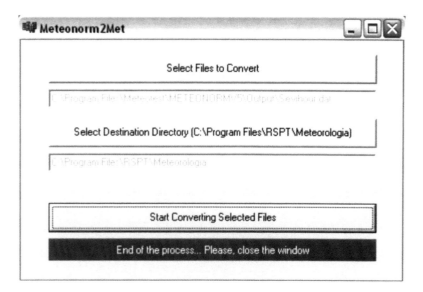

Meteonorm2Met

Select Files to Convert

Select Destination Directory (C:\Program Files\RSPT\Meteorologia)

Start Converting Selected Files

Fig. B.20   Input screen of the Meteonorm2Met utility.

Meteonorm2Met

Select Files to Convert

C:\Program Files\Meteotest\METEONORM V5\Output\Sevihour.dat

Select Destination Directory (C:\Program Files\RSPT\Meteorologia)

C:\Program Files\RSPT\Meteorologia

Start Converting Selected Files

End of the process... Please, close the window

Fig. B.21   Output screen of the Meteonorm2Met utility.

INDEX

T - #0918 - 101024 - C172 - 210/279/7 - PB - 9781844073139 - Gloss Lamination